T0315046

LIQUID CHROMATOGRAPHY TIME-OF-FLIGHT MASS SPECTROMETRY

CHEMICAL ANALYSIS

A SERIES OF MONOGRAPHS ON ANALYTICAL CHEMISTRY AND ITS APPLICATIONS

Series Editor
J. D. WINEFORDNER

Volume 173

A complete list of the titles in this series appears at the end of this volume.

LIQUID CHROMATOGRAPHY TIME-OF-FLIGHT MASS SPECTROMETRY

Principles, Tools, and Applications for Accurate Mass Analysis

EDITED BY
IMMA FERRER
E. MICHAEL THURMAN

WILEY

A JOHN WILEY & SONS, INC., PUBLICATION

Published by John Wiley & Sons, Inc., Hoboken, New Jersey.
Published simultaneously in Canada.

For general information on our other products and services or for technical support, please contact our
Customer Care Department within the United States at (800) 762-2974, outside the United States at
(317) 572-3993 or fax (317) 572-4002.

Wiley also publishes its books in a variety of electronic formats. Some content that appears in print
may not be available in electronic formats. For more information about Wiley products, visit our web
site at www.wiley.com.

Library of Congress Cataloging-in-Publication Data:
Ferrer, Imma, 1970–
      Liquid chromatography time-of-flight mass spectrometry : principles, tools, and applications for
accurate mass analysis / Imma Ferrer, E.M. Thurman.
          p.   cm.
      Includes index.
      ISBN 978-0-470-13797-0 (cloth)
  1. Liquid chromatography.   2. Time-of-flight mass spectrometry.   I. Thurman, E. M.
(Earl Michael), 1946–   II. Title.
      QD79.C454.F47 2008
      543′.84—dc22

                                                                          2008032176

10 9 8 7 6 5 4 3 2 1

CONTENTS

PREFACE

"Most of the knowledge and much of the genius of a researcher lie behind ones selection of a problem worth doing."
—*Alan Gregg of the Rockefeller Foundation*

We finished our first book, *Liquid Chromatography/Mass Spectrometry, MS/ MS, and Time-of-Flight MS: Analysis of Emerging Contaminants*, six years ago now. Since then the subject of accurate mass analysis has increased exponentially from less than 100 papers to over a thousand. Instruments, too, have changed during this time. We have witnessed an increase in mass accuracy from greater than 5 ppm to less than 1 ppm! Resolving power has increased from 2000 to 5000 for benchtop TOF instruments to now routine 10,000 to 20,000. New accurate mass spectrometers have been developed, such as the Orbitrap, which is capable of resolving power of nearly 100,000, which is comparable to Fourier transform ion cyclotron mass spectrometers. Thus, we felt it was time to assemble a group of seasoned authors to write chapters on accurate mass analysis, including the fundamentals of accurate mass by liquid chromatography/time-of-flight mass spectrometry, the diverse tools of TOF such as exact-mass databases, isotopic mass defects, isotopic ratios and neutral losses, and the subject of accurate mass of fragment ions. We also have applications using TOF to identify unknowns and environmental problem solving for pesticides, pharmaceuticals, hormones, and natural products in food and water.

In the introduction of our last book on LC/MS/MS and LC/TOF-MS, we stated a group of future trends. Lets see if any of those trends have come true. First was the development of a searchable library for LC/MS and LC/MS/MS triple quadrupole. This prediction is coming true not for the triple quadrupole, but instead for LC/TOF-MS. The ability of TOF to give the elemental composition of a compound allows for the development of exact-mass databases. This idea plays an important role in several chapters in this book. The second prediction was the increased use of LC/TOF-MS and LC/Q/TOF-MS for elemental composition of unknowns and fragment ions. This is exactly what has happened. Furthermore, we see that LC/TOF-MS plays an important role here even without the ability to do MS/MS. This is because of the ability to do in-source CID (collision induced dissociation) and to use deconvolution software to unravel the fragment ions and align them with their respective precursor ion.

The third prediction was the routine use of LC/MS/MS triple quadrupole for routine monitoring. This is definitely true and we see that the LC/MS single quadrupole has its days numbered to be replaced by the triple quadrupole. Whereas the

LC/TOF-MS is staking out new ground separated from the LC/MS/MS triple quadrupole. On the other hand, the LC/Q/TOF-MS has not taken a large share of the environmental market away from the LC/TOF-MS. The fourth prediction was the direct analysis of water by triple quadrupole analysis. This prediction is still playing out as we see publication of more methods by LC/MS/MS triple quadrupole on water samples directly, especially as instrument limits of detection continue to go to lower and lower values. The LC/TOF-MS is competing with the triple quadrupole in the area of screening large numbers of samples; in fact one of the chapters in this book discusses the "crossover point" or the point where the limits of detection of both instruments are nearly equal.

But something new has entered the fray of direct sample analysis. It is the direct analysis of surfaces by two new methods of ionization called "DART" and "DESI." These are both surface analysis methods using a similar ionization to electrospray to generate ions for subsequent analysis by LC/TOF-MS or LC/MS/MS. Several chapters address this new method of ionization for direct analysis of food and pharmaceuticals, something that was not imagined six years ago. The fifth prediction was that libraries of characteristic ions would be developed. We see this coming true, but at a slower pace than we might desire. New software is being developed to take full advantage of the power of sub 1-ppm mass accuracy of both precursor and product ions. The sixth and last prediction was the marriage of trap and TOF into a single instrument. Although this was accomplished in two ways, one by trap and TOF (Shimadzu) but also by trap and Orbitrap (Thermo Scientific), the expense of these instruments has nearly prevented their use in environmental analysis. So our future predictions have been near the mark, with a few surprises thrown in. Thus, we can expect similar exciting things in the next 5 to 10 years.

As we look into the future of LC/TOF-MS, what we see on the horizon are the following:

1. Large and universal databases for easy use via the internet. This would include software tools to help in the identification of true unknowns. A true unknown being a compound not in a database but identifiable via exact mass measurements of it and its fragment ions.

2. Next is a continued increase in resolving power, which is driven by the invention of a competitive instrument, the Orbitrap that has raised the bar of resolving power to 50,000 or more. Accuracy will be more difficult to improve on but will hover around the 1-ppm range.

3. The increased and routine use of LC/TOF-MS in the environmental market with retrospective analysis. This is the ability to return to old data files to discover the identity of unknown compounds and their degradates or metabolites.

4. The size of data files has increased from 50 megabyte files to 400 megabytes. As the resolving power continues to increase so does the memory capacity.

5. Fast chromatography will also play a role in accurate mass analysis. We predict that it will drive the use of LC/TOF-MS since it has the ability to do accurate analysis on 2 second peaks with not only good accuracy but with 5 to

10 scans per peak for excellent chromatographic resolution and excellent mass measurement.

This book compiles the work of many authors who are pioneering the advent of time-of-flight techniques applied to environmental problems. Interestingly enough we came up with a group of scientists who use a diversity of commercially available accurate mass instruments as well. We thought it was worthy to mention the commercial companies that make these analyses possible. In this book, many interesting examples are given for time-of-flight mass spectrometers from (in alphabetical order): Agilent, Bruker Daltonics, Jeol, and Waters.

In summary, we thank all of our authors who have contributed their time, intellect, and effort to this new book. Thank you to everyone.

Imma Ferrer
Mike Thurman

Longmont, Colorado
April 2009

CONTRIBUTORS

Facundo M. Fernández, Christina Y. Hampton, Leonard Nyadong, Arti Navare, and Mark Kwasnik
Assistant Professor, School of Chemistry and Biochemistry
Georgia Institute of Technology
Atlanta, GA

Imma Ferrer
Research Scientist, Center for Environmental Mass Spectrometry (CEMS)
Department of Civil, Environmental and Architectural Engineering
University of Colorado
Boulder, CO

John C. Fjeldsted
Director, Research and Development, LC/MS Division
Agilent Technologies, Inc.
Santa Clara, CA

Jeffrey R. Gilbert, Jesse L. Balcer, Scott A. Young, Dan A. Markham, Dennis O. Duebelbeis, and Paul Lewer
Technical Leader, Mass Spectrometry CofE
Dow AgroSciences
Indianapolis, IN

Andrew H. Grange and G. Wayne Sovocool
Research Scientist, Environmental Chemistry Branch
U.S. EPA, ORD, NERL, Environmental Sciences Division
Las Vegas, NV

Félix Hernández, Juan V. Sancho, and María Ibáñez
Research Scientist, Research Institute for Pesticides and Water
University Jaume I
Castellón, Spain

Anton Kaufmann
Research Scientist, Official Food Control Authority of the Canton of Zurich
Kantonales Labor Zürich
Zürich, Switzerland

Ilkka Ojanperä, Anna Pelander, and Suvi Ojanperä
Laboratory Director, Department of Forensic Medicine
University of Helsinki
Helsinki, Finland

O. David Sparkman, Patrick R. Jones, and Matthew Curtis
Professor, Pacific Mass Spectrometry Facility, Department of Chemistry
College of the Pacific, University of the Pacific
Stockton, CA

Alida (Linda) A.M. Stolker
Research Scientist, Analysis and Research/Veterinary Drugs
RIKILT- Institute of Food Safety
Wageningen, The Netherlands

E. Michael Thurman, Jerry A. Zweigenbaum, and Paul A. Zavitsanos
Director, Center for Environmental Mass Spectrometry (CEMS)
Department of Civil, Environmental and Architectural Engineering
University of Colorado
Boulder, CO

Michael C. Zumwalt
LC-MS Product Specialist, LC&LC/MS Solutions
Agilent Technologies, Inc.
Englewood, CO

PART 1

PRINCIPLES AND THEORETICAL ASPECTS OF ACCURATE MASS

ACCURATE MASS MEASUREMENTS WITH ORTHOGONAL AXIS TIME-OF-FLIGHT MASS SPECTROMETRY

John C. Fjeldsted

Director, Research and Development, LC/MS Division, Agilent Technologies, Inc.,
Santa Clara, California

AN APPEALING attribute of time-of-flight mass spectrometry is its fundamental simplicity. It is easy to conceptualize that for a given population of ions all accelerated to the same energy those with the lightest mass will travel with a greater velocity and reach an end point sooner. Simply recording these arrival times and providing a calibration function converts ion arrival times to mass values and produces a simple spectrum.

From simple beginnings time-of-flight has transformed itself into a very powerful analytical tool. The success of orthogonal axis time-of-flight mass spectrometry is the result of both fundamental improvements in analyzer geometry and design as well as technological advances such as those offered by ultra-high-speed signal digitization systems.

This chapter is aimed at giving the practitioner a basic understanding of the underlying theory of TOF mass analysis and instrumental factors critical to achieving the high performance required to meet today's demanding applications.

1.1 INTRODUCTION

Since the introduction of commercial orthogonal axis–time-of-flight (oa-TOF) instruments in the mid-1990s, significant improvements have been made to both atmospheric sampling ion sources and TOF mass analysis. Today several manufacturers offer MS and MS/MS instruments which take advantage of oa-TOF's unique combination of speed, sensitivity, resolving power, and mass accuracy. In

Liquid Chromatography Time-of-Flight Mass Spectrometry: Principles, Tools, and Applications for Accurate Mass Analysis, Edited by Imma Ferrer and E. Michael Thurman
Copyright © 2009 John Wiley & Sons, Inc.

combination with a liquid chromatograph oa-TOF is a power tool for the analysis and identification of trace organic substances.

1.1.1 History of Time-of-Flight Mass Analysis

The origin of time-of-flight mass spectrometry dates back to 1946 when Stephens presented the concept at the American Physical Society [1] and was first demonstrated by Cameron and Eggers with the analysis of mercury vapor [2]. An important advance in resolving power was achieved by Wiley and McLaren [3] in 1955 with the introduction of space focusing. With additional improvements in ion detection, resolving powers up to 300 where achievable. In 1958 the first commercial TOF instrument was manufactured by Bendix [4]. Due to the high-speed measurement associated with TOF, spectra were recorded with electronic oscilloscopes.

The next critical development needed for TOF was a means of overcoming the effects of initial ion energy spread on flight time. In 1966 Mamyrin, in his doctorate thesis [5], put forth the concept of refocusing ion energy spread which he confirmed by measurement in 1971 [6].

The development of matrix-assisted laser desorption ionization (MALDI) brought a resurgence in TOF mass analysis in the 1980s and 1990s. In contrast to pulsed ionization development, effective coupling to continuous ion sources was first proposed by Dawson and Guilhaus [7, 8] in 1989 and reported by Dodonov et al. in 1991 [9]. The resulting instrumental design is referred to as orthogonal axis time-of-flight mass spectrometry (oa-TOF) and has steadily grown in use due to the wide spread adoption of atmospheric pressure ionization and in particular electrospray ionization. See references [10, 11] for in depth history and instrumental details.

1.2 THEORY OF OPERATION

1.2.1 Equations for Time-of-Flight

The flight time for each ion of particular m/z is unique. The flight time begins when a high-voltage pulse is applied to the back plate of the ion pulser (see Figure 1.1) and ends when the ions of interest strike the detector. The flight time (t) is established by the energy (E) to which an ion is accelerated, the distance (d) it has to travel, and its mass (strictly speaking its mass-to-charge ratio).

There are two well-known formulae that apply to time-of-flight analysis. One is the formula for kinetic energy, the energy of an object (or an ion) in motion, which is expressed as:

$$E = \frac{1}{2}mv^2$$

which solved for m becomes:

$$m = 2E/v^2$$

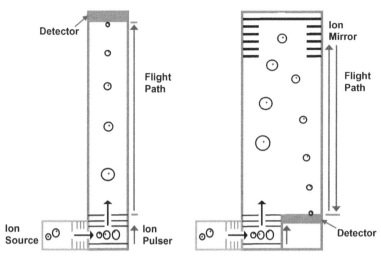

Figure 1.1.   Orthogonal axis time-of-flight mass spectrometers with the configuration on the left showing linear geometry and the configuration on the right incorporating an ion mirror.

The second equation is the familiar equation where velocity (v) equals distance (d) divided by time (t) or:

$$v = d/t$$

Combining the first and second equations yields:

$$m = (2E/d^2)t^2$$

This gives us the basic time-of-flight relationship. For a given energy (E) and distance (d) the mass is proportional to the square of the ion flight time.

The equation stipulates that for a given kinetic energy, E, smaller masses will have larger velocities, and larger masses will have smaller velocities. Hence, ions with lower masses arrive at the detector earlier, as shown in Figure 1.1.

In the design of an oa-TOF mass spectrometer, much effort is devoted to holding the values of the energy (determined by the high voltages) and the distance the ion travels constant, so that an accurate measurement of flight time will give an accurate mass value. As these terms are held constant they are often combined into a single variable, A, so:

$$m = A(t)^2$$

This is an ideal equation, based on true flight times. In practice, there is a delay from the time the control electronics send a start pulse to the time that high voltage is actually present on the ion pulser plates. There is also a delay from the time an ion reaches the front surface of the ion detector until the signal generated by that ion is digitized by the acquisition electronics. These delays are very short, but not insignificant. Because the true flight time cannot be measured, it is necessary to correct the measured time, t_m, by subtracting the sum of both the start and stop delay times which, when added together, are referred to as t_o.

$$t = t_m - t_o$$

By substitution, the basic formula that can be applied for actual measurements becomes:

$$m = A(t_m - t_o)^2$$

1.2.2 Mass Calibration

To make the conversion from measured flight time, t_m, to mass, the values of A and t_o must be determined, so a calibration is performed. A mixture of compounds whose exact masses are known with great accuracy is analyzed. Then a simple table (Table 1.1) is established of the flight times and corresponding known masses.

Now that m and t_m are known for a number of values across the mass range, the computer that is receiving data from the instrument does the calculations to determine A and t_o. Using an intelligent algorithm, it tries different values of A and t_o until the right side of the calibration equation,

$$m = A(t_m - t_o)^2$$

matches as closely as possible the left side of the equation (m), for all seven of the mass values in the calibration mix.

While this initial determination of A and t_o is highly accurate, it is still not accurate enough to give the best possible mass accuracy for time-of-flight analysis. A second calibration step is needed. After the calibration coefficients A and t_o have been determined, a comparison is made between the actual mass values for the calibration masses and their calculated values from the equation. These typically deviate by only a few parts-per-million (ppm). Because these deviations are small and relatively constant over time, it is possible to perform a second-pass correction to achieve an even better mass calibration. This is done via an equation that corrects the small deviations across the entire mass range. This correction equation (typically a higher-order polynomial function) is stored as part of the instrument calibration. The remaining mass error after this two-step calibration method, neglecting all other instrumental factors, is typically at or below 1 ppm over the range of calibration masses.

TABLE 1.1. TOF Mass Calibration

Calibrant Compound Mass (m)	Flight Time (μsec) (t_m)
118.0863	20.79841
322.0481	33.53829
622.029	46.12659
922.0098	55.88826
1521.971	71.45158
2121.933	84.14302
2721.895	95.13425

1.2.3   TOF Measurement Cycle

TOF measurements do not rely on the arrival times of ions coming from just a single pulse applied to the ion pulser, but instead are summations of the signals resulting from many pulses. Each time a high voltage is triggered to the plates of the ion pulser, a new spectrum (called a single transient) is recorded by the data acquisition system. This is added to previous transients until a predetermined number of sums have been made. For analyses requiring a scan speed of one spectrum per second, approximately 10,000 transients can be summed before transferring the data from the instrument back to the host computer to be written to disk. If the target application involves high-speed chromatography, then fewer transients are summed and the rate at which spectra are recorded is increased.

The mass range limits the number of times per second that the ion pulser can be triggered and transients recorded. Once the ion pulser fires, it is necessary to wait until the last mass of interest arrives at the ion detector before the ion pulser is triggered again. Otherwise light ions triggered from the second transient would arrive before the heavier ions of the first transient, resulting in overlapping spectra.

Table 1.2 shows several example masses with their approximate flight times and possible transients/second. These are calculated for an effective flight length of 2 meters and an accelerating potential of 6500 volts. Under these conditions, a mass of 3200 has a flight time of about 100 microseconds (μsec). As there is essentially no delay time between transients, this means that 10,000 transients/second correspond to a mass range of $3200 \, m/z$. For a smaller mass range, the ion pulser can be triggered at higher rates. For example, a mass range of $800 \, m/z$ (one-fourth of $3200 \, m/z$) reduces the flight time to 0.1 msec/$\sqrt{4}$, or 0.05 milliseconds, allowing for 20,000 transients/second. Conversely, extending the transient to 0.141 milliseconds doubles the mass range to $6400 \, m/z$ (remember, mass is a function of the time squared).

Because each transient takes place in such a short period of time, the number of ions for any particular compound at a specific mass is often quite small and for many oa-TOF instruments traditionally substantially less than one. Historically, this fact plays an important role in the basic design of the data acquisition system of many of today's commercial instruments.

TABLE 1.2.   Flight Time and Transients/Second as a Function of Mass*

m/z	Flight Time (μsec)	Transients/sec
800	50	20,000
3200	100	10,000
6400	141	7070

*Two-meter flight effective flight path, flight potential 6500 V.

1.3 THE RELATIONSHIP BETWEEN MASS RESOLVING POWER AND MASS ACCURACY

1.3.1 Instrumental Limitation to Mass Resolving Power

Even in the case of an ideal mass spectrometer which exhibited perfect ion acceleration and flight distance stability there exists factors that produce variations in the ion arrival time for a given mass. The major source of variation arises from the position and energy each ion possess in the ion pulser at the instant that the high-voltage pulse initiates each transient measurement cycle.

The first of these two potential sources of variation, position in the ion pulser, is largely corrected for using Wiley-McLaren space focusing [3, 12]. Under Wiley-McLaren space focusing, a potential gradient is established across the ion pulser region. With a higher potential at the rear of the pulser region, these ions which have a greater distance to travel also receive a greater acceleration potential. The result is that the ions accelerated from the rear of the ion pulser actually "catch up" with those closer to the front of the pulser and are therefore refocused in time.

The second source for variation in ion arrival time is related to the initial energy of each ion in the ion pulser and is commonly referred to as the "turnaround time." To understand this effect consider two ions of equal mass and both at the same identical position in the ion pulser. The first of these two ions has a small residual energy in line with the field that the pulser applies to accelerate the ions for the time-of-flight measurement. The second has the same small residual energy, but its velocity is directed against the orthogonal accelerating field. At the instant the high-voltage pulser is triggered the first ion accelerates quickly in the desired direction. For a brief moment, the second ion travels in the opposite direction until its initial energy is lost, then re-accelerates past its origin, delayed from the launch of the first ion by a short period of time. The turn around time, Δt, can be calculated by the expression

$$\Delta t = 2v/a$$

where v is initial ion velocity directed against the desired orthogonal flight path and a is the ion acceleration in pulser field.

There are three principal approaches to minimize the broadening of ion arrival time. The first is to minimize the energy spread of the beam entering the ion pulser. Most instruments take advantage of beam forming optics and slits to minimize beam divergence entering the pulser. For clarification, it is the energy of the ions orthogonal to entry into the pulser (i.e., the axis of the time of flight measurement) that is critical.

The second instrumental factor is to increase the potential applied to the plates that accelerate the ions in the ion pulser thereby increasing the accelerating field. This plays together with the instrumental requirements for Wily-McLaren space focusing. In practice it also creates a rather large spread in ion energy for ions exiting the ion pulser. It is for this reason that all oa-TOF instruments make use of an ion mirror that compensates for this energy spread.

The reflectron or Mamyrin ion mirror [6] is placed at the end of the flight tube where in "linear" instruments the ion detector is positioned. The mirror is constructed with a stacked set of metal rings each having a higher electrical potential applied. Ions penetrate the field established by these rings until they loose all forward energy and their direction of travel is reversed. The greater the energy the ions possess entering the ion mirror, the greater the penetration and resulting distance that is traveled. The potential field in the ion mirror is precisely established so that the time required to travel the additional distance is exactly canceled by the ion's greater ion energy.

The third instrumental factor is to increase the flight time of the measured ions. Because mass is proportional to the square of ion arrival time the mass resolving power is one half the ion arrival time (t) divided by the spread in ion arrival time (Δt), which is principally the ion turnaround time:

$$\text{Mass Resolving Power} = t/(2\Delta t)$$

For a given acceleration potential the ion arrival time, and hence the mass resolving power, is established by the length of the ion flight path. With a 2 meter effective flight path and a corresponding flight time of 50 µsec a turnaround time of 1.67 nsec results in a resolving power of 15,000 using the mass resolving power equation above.

1.3.2   The Effect of Ion Detection on Resolving Power

In an ideal instrument the exact arrival time for each ion would be detected and recorded without any effect on the actual ion population being measured. In reality, with high ion currents and an arrival time variation on the order of 1 nsec this becomes very challenging to achieve instrumentally. While detector response times can reach down to the 100s of picoseconds, this is still insufficient with high-sensitivity designs that can present up to 100s of ions within this 1 nsec spread. Under such conditions, detecting individual ion arrival times is not possible. To get around this challenge two different approaches have been developed.

The historical approach for time-of-flight ion arrival detection is based on a Time to Digital Converter (TDC) together with a discriminator to selectively register an arrival when the output of the ion detector crosses a set threshold. This approach was considered important so as to maximize ion detection and reduce electronic noise. The limitation when using such a threshold detection system is that, as sample levels increase, multiple ion arrivals occur for a given mass within a single transient. When such is the case two consequences occur. The first is that the earliest ion to arrive for a given mass triggers the TDC with some of the later part of the distribution being missed. This results in an apparent shift towards lower arrival times and hence a shift towards lower reported mass. As ion intensity continues to increase, individual ions have coinciding arrival times which results in only a single response, thereby truncating the recorded signal. This causes an amplitude saturation effect, and for this reason, oa-TOF has historically not been considered suitable for quantitative analysis.

One approach found in commercial instrumentation to extended dynamic range is through the use of a segmented detector and a multi-input TDC. By reduction of the ion capture area of an individual segment, a higher signal level can be accommodated.

More recently, some instruments that use TDC acquisition systems have incorporated controllable beam attenuator which can be automatically invoked by the instrument acquisition system to decrease the ion transmission before it is pulsed for mass measurement. This results in an increase of dynamic range, but at the cost of losing ion signal which ultimately limits mass accuracy and reduces in-scan dynamic range.

A more recent approach to recording ion arrival time for oa-TOF instruments makes use of high-speed Analog to Digital Converter (ADC) technology. Unlike the discriminator TDC approach, an ADC is able to measure the analog response of the ion detector and can track the ion abundance as it increases. This gives ADC-based systems greater intrinsic dynamic range. The downside for ADC-based systems can be found in the fact that included in the ion arrival measurement is an apparent broadening of the ion arrival distribution. This can be explained when one considers that the analog response of the ion detector has a finite width which gets summed into digitization process.

1.3.3 Translating Mass Resolving Power to Mass Accuracy

At this point we have determined a nominal mass resolving power for an oa-TOF mass spectrometer. A resolving power specification of 15,000 is shown for m/z 1500 in Figure 1.2. A simple calculation gives us the peak width (FWHM) of this mass peak of 0.10 m/z.

If the mass measurement obtained from the mass spectrometer pertains only to that of single ions then the mass accuracy is directly related to the resolving power. As measurement accuracy is generally specified using standard deviation, conver-

$$\text{FWHM} = 2\sqrt{2\ln(2)}\ \sigma \sim 2.35\,\sigma$$

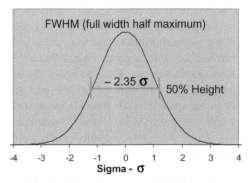

Figure 1.2.   Relationship between FWHM and standard deviation for a normal distribution.

sion between a peak's full width half maximum value and standard deviation is required.

Considering a mass at m/z 1500 and a resolving power of 15,000 the observed FWHM peak width is 0.1 m/z. The standard deviation for individual ion arrival (1σ) is equal to 0.043 m/z. It is customary to specify the mass accuracy as a relative unit expressed in parts per million, or in this case 28 ppm (1σ). As $\pm 1\sigma$ represents only 68.3% of the measurement distribution, 2σ values representing 95.4% of the measurement distribution are often specified.

Absent from the above calculations of mass measurement accuracy is the consideration that each mass measurement is the summation of a large number of transients and generally a large number of ions which populate the arrival distribution. The relationship between the number of ions contributing to the distribution reduces the uncertainty when establishing the mean. This improvement in mass measurement accuracy follows the central limit theorem according to

$$\sigma_{mean} = \sigma_{single\ event} / \sqrt{n}$$

where n = number of single events included in the mean.

Continuing the previous example, the distribution for a single ion arrival with a resolving power of 15,000 has a 2σ confidence interval of 56 ppm, the accuracy of assigning the mean is reduced by the square root of the number of ions detected (see Table 1.3).

The number of ions detected in a measurement is dependent on the quantity of analyte and the sensitivity of the mass spectrometer. While mass spectrometers vary in ionization and transmission efficiency, for standard electrospray operation it is estimated that 1 ng of a reasonably well ionizing analyte results in between 10,000 and 100,000 detected ions depending on the instrument. For fragment ions of an MS/MS spectrum this value is typically reduced by the fraction of ions associated in the MS/MS transition of interest. When based solely on resolving power and ion statistics, oa-TOF can achieve 2 ppm mass accuracy with as little as 10 pg when measured with the most sensitive of instruments.

TABLE 1.3. Relationship Between Ion Statistics and
Theoretical Limit of Mass Measurement Accuracy for a
Resolving Power of 15,000

Number of Ions Detected	2 sigma (95% confidence)
1	56 ppm
10	18 ppm
100	5.6 ppm
1000	1.8 ppm
10,000	0.56 ppm
100,000	0.18 ppm

1.4 OPERATIONAL FACTORS AND PRACTICAL LIMITS

1.4.1 Reference Mass Correction

Achieving an accurate mass calibration is the first step in producing accurate mass measurements. When the goal is to achieve accuracies at the 1 ppm level, even the most miniscule changes in accelerating potentials or flight distances can cause a noticeable shift. It is possible to cancel out these instrumental drift factors with the use of reference mass recalibration. With this technique, compounds of known mass are introduced into the ion source while samples are being analyzed. The presence of reference masses in the acquired spectrum allows calibration corrections to be made on individual or averaged mass spectra. This can be accomplished by the data processing software, or ideally it is performed automatically during the analysis as spectra are initially stored to disk.

1.4.2 Mass Assignment Errors Due to Chemical Interferences

The second significant factor that limits mass accuracy is chemical background. The high resolving power of a TOF system helps to reduce the chances of having the peak of interest merged with background, yet even a small unresolved impurity can shift the centroid of the expected mass. The magnitude of this effect can be estimated by using a simple weighted average calculation:

$$\Delta_{obs} = \Delta_{contaminant} \cdot Abd_{contaminant} / (Abd_{contaminant} + Abd_{sample})$$

where

Δ_{obs} is the observed shift in mass in ppm

$\Delta_{contaminant}$ is the mass difference between the sample and contaminant in ppm

$Abd_{contaminant}$ and Abd_{sample} are the mass peak heights or areas of the contaminant and sample

For a resolving power of 10,000, a mass difference between the sample and contaminant of 50 ppm, and relative mass peak heights of 10:1 (sample vs background) the observed mass shift would be 50 · 1 / (1 + 10) or about 5 ppm.

There are a number of ways to minimize chemical background. Of instrumental importance is to use system with a sealed ion source chamber that minimizes contamination from the laboratory air. It is also essential to use high-purity HPLC solvents and follow a systematic cleaning program for the HPLC and the ion source.

1.4.3 Enhanced Performance Through Novel Acquisition Systems

The combination of ultra-high-speed sampling and real-time signal processing enables two new modes of operation. They are referred to as Transient Level Peak Picking (TLPP) and simultaneous dual gain mode. In TLPP mode, as each TOF

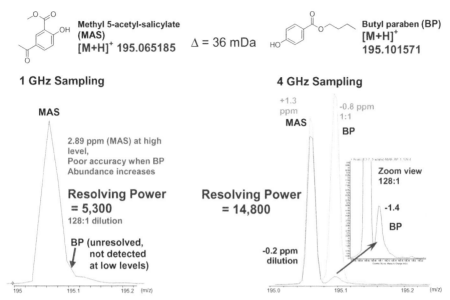

Figure 1.3.   Comparison of resolving power achievable with 1 GHz and advanced 4 GHz technology.

transient (a transient is the individual time-of-flight spectrum acquired from each TOF pulse) is acquired the precise arrival time of the ion packet for each mass is determined in real time. This maximizes resolution which is similar to operation with a TDC. What differentiates TLPP from a TDC is that the arrival time of each packet of ions for each mass is triggered at the peak apex instead of the first ion or leading edge of the ion packet. Using apex detection not only is the arrival time recorded, but also the signal intensity. This is only possible with an ADC and dramatically extends the in-spectrum dynamic range. Apex detection also eliminates the need for dead time correction critical to TDC operation.

In dual gain mode the 4 GHz sampling rate is divided into two channels of data acquisition. One channel operates with the standard preamplifier gain. The second channel operates at a reduced gain. Both channels are acquired simultaneously at 2 GHz. In this way there is never a need to attenuate the ion current transferred to the TOF mass spectrometer. Once the desired number of TOF transients has been summed, the standard and reduced gain spectra are merged in such a way that any peaks that saturated the standard gain channel are replaced by the data coming from the reduced gain channel. The advantage of dual gain mode is that all ions are detected, maximizing ion statistics, which improves mass accuracy while simultaneously extending the dynamic range.

With 4 GHz sampling and TLPP higher resolving power is now possible with ADC systems. Figure 1.3 shows how two isobars can now be fully resolved even with a broad spread in concentration.

With dual gain mode, spectra are simultaneously acquired at standard and at low gain levels. Before spectra are sent from the mass spectrometer to the host

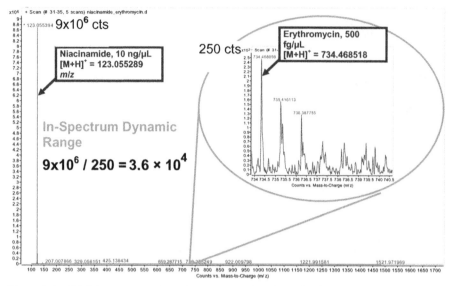

Figure 1.4. High in-spectrum dynamic range resulting from simultaneous dual gain acquisition.

computer, any mass peaks in the spectrum which are saturated at the standard gain spectrum are replaced with corresponding data points from the low gain spectrum. Figure 1.4 shows the detection of both high and very low level analytes differing in response by 4.5 decades.

1.5   CONCLUSIONS

What sets oa-TOF apart from other methods of mass analysis is its unique combination of speed, mass accuracy, sensitivity and dynamic range. Table 1.4 indicates performance achievable on leading instruments. What is unique to TOF is that each of these specifications can be simultaneously achieved when working with significant signal to noise and ion statistics.

As compared to quadrupole mass filter-based mass spectrometers, oa-TOF offers higher mass accuracy, resolving power, full-spectrum sensitivity, and speed, while quadrupole mass filters offer the ultimate in single ion monitoring sensitivity and instrumentation cost.

As compared to 2D and 3D ion trap mass spectrometers, oa-TOF has greater mass accuracy, resolving power, speed, and, with the latest generation acquisition systems, in-spectrum dynamic range. Ion traps on the other hand generally have the highest full scan sensitivity.

While FT-ICR systems demonstrate unsurpassed resolving power at low *m/z* ions and low data acquisition rates, other key performance factors favor TOF. These include in-spectrum dynamic range, speed and cost. When considering mass accu-

TABLE 1.4. Properties of a High Performance oa-TOF System

Property	Specification
Mass accuracy	<2 ppm
Mass range	>10,000 m/z
Resolving power	Up to 20,000
Sensitivity	<10 pg
Dynamic range—in spectrum	Up to 5 decades
Speed—spectra/sec	Up to 40/sec
Cost	Moderate

racy the difference between nonmagnetic Fourier transform (Orbitrap) and high-performance oa-TOF systems appears to be less than a factor of 2.

Considering the complexity of many current LC/MS applications, oa-TOF offers key analytical value particularly in light of the ability to effectively couple to a high-resolution chromatographic separation system, where high speed, high dynamic range, and very low part-per-million mass accuracy is required.

REFERENCES

1. Stephens, W.E., A pulsed mass spectrometer with time dispersion, *Phys. Rev.*, **1946**, 69, 691.
2. Cameron, A.E.; Eggers, D.F., An ion "velocitron," *Rev. Sci. Instrum.*, **1948**, 19, 605–607.
3. Wiley, W.C.; McLaren, I.H., Time-of-flight mass spectrometer with improved resolution, *Rev. Sci. Instrum.*, **1955**, 26, 1150–1157.
4. Scripps Center for Mass Spectrometry website, http://masspec.scrips.edu/mshistory/perspectives/timeline.php
5. Mamyrin, B.A., Time-of-flight mass spectrometry (concepts, achievements, and prospects), *Int. J. Mass Spectrom.*, **2001**, 206, 251–266.
6. Karataev, V.I.; Shmikk, D.V.; Mamyrin, B.A., New method for focusing ion bunches in time-of-flight mass spectrometers, *J. Tech. Phys.*, **1972**, 16, 1177–1189.
7. Dawson, J.H.J.; Guilhaus, M., Orthogonal-acceleration time-of flight mass spectrometer, *Rapid Commun. Mass Spectrom.*, **1989**, 3, 155–159.
8. Coles, J.; Guilhaus, M., Orthogonal acceleration—a new direction for time-of-flight mass spectrometry: fast, sensitive mass analysis for continuous ion sources, *Trends Anal. Chem.*, **1993**, 12, 203–213.
9. Dodonov, A.F.; Chernushevich, I.V.; Laiko, V.V., *Proceedings of the 12th International Mass Spectrometry Conference,* 26–30.
10. Igor, V.; Chernushevich, A.; Loboda, V.; Thomson, B.A., An introduction to quadrupole time-of-flight mass spectrometry, *J. Mass Spectrom.*, **2001**, 36, 849–865.
11. Tamura, J.; Osuga, J., Next Generation LC-TOF/MS—AccuTOF, Application Note, JEOL, Ltd., 2001.
12. Seccombe, D.P.; Reddish, T.J., Theoretical study of space focusing in linear time-of-flight mass spectrometers, *Rev. Sci. Instrum.*, **2001**, 72, 1330–1338.

CHAPTER 2

THE MASS DEFECT, ISOTOPE CLUSTERS, AND ACCURATE MASS FOR ELEMENTAL DETERMINATION

E. Michael Thurman and Imma Ferrer

Center for Environmental Mass Spectrometry, Department of Civil, Environmental and Architectural Engineering, University of Colorado, Boulder, Colorado

T HE POWER of accurate mass is in the determination of the elemental composition of a compound or its fragment ions. Three aspects of the measurement of accurate mass lead to a correct formula. They are mass accuracy, the mass defect of the protonated or de-protonated molecule, and the accuracy and intensity of the isotopic composition pattern. Used together these three types of information are valuable to obtain the correct formula of an unknown. With the correct formula it is possible to compare this to a database for unknown identification, which is a topic discussed later in this book (Chapter 4, Ferrer and Thurman). The first aspect is mass accuracy expressed in either ppm or millimass units. The number of possible formulae increases exponentially as the molecular mass of a compound increases, while the number of formulae go up only arithmetically as the ppm error increases. Thus, a 1 ppm error at mass 200 has only 1 or 2 formulae possible while a 1 ppm error at mass 1000 has 20 formulae. The factors affecting accuracy typically include the calibration of the instrument and the calibration curve itself. The second aspect is the mass defect, which is the difference in mass between an element and its nominal mass. Mass defects that are positive reflect the incorporation of hydrogen in the molecule, while large negative mass defects suggest atoms such as S, Cl, and other halogens. Lastly, the isotopic composition pattern is extremely valuable as a check of the formula and its correctness. This chapter discusses these three topics and how to use them effectively for the determination of molecular formula and the identity of unknown compounds.

Liquid Chromatography Time-of-Flight Mass Spectrometry: Principles, Tools, and Applications for Accurate Mass Analysis, Edited by Imma Ferrer and E. Michael Thurman
Copyright © 2009 John Wiley & Sons, Inc.

2.1   INTRODUCTION

The identification of unknowns by accurate mass (e.g., LC/TOF-MS) requires four steps, beginning with the identification of the correct molecular formula [1]. This first step is critical in the analysis of an unknown. Second is to find possible candidate structures for the molecular formula using a database approach [2]. Third is to then check the candidate structures against product ions that were formed by fragmentation of the parent compound. And fourth or lastly is to confirm the identification by actual standard analysis, confirming both the retention time and products ions by accurate mass [1]. This procedure has been successfully applied to the analysis of pesticides in food [1], pharmaceuticals in water and sediment [3], surfactants in wastewater [4], and other environmental applications [5, 6].

Because both step 1 and step 3 require the determination of a molecular formula, any tools that can help in the correct identification of the molecular formula are important and useful, especially with respect to A + 2 elements, such as Cl, Br, and S. This chapter describes such new tools, the relative isotopic mass defect (RΔm), and the calculation of the isotopic mass average (IMA), which are used in the identification of the correct formula of the protonated (or de-protonated) molecule and any of its product ions. This chapter explains the conceptual tool and how it is used along with the intensity of the isotope cluster and a formula generator to define the least number of molecular formulas for an unknown compound. There are examples of pesticides in food and water, which will show the application of the relative isotopic mass defect and the isotope mass average (IMA).

2.2   THEORETICAL CONSIDERATIONS

2.2.1   The Physicist's Mass Defect

Careful mass measurements of the elements have shown that the mass of a particular atom is always slightly less than the sum of the masses of the individual protons, neutrons, and electrons of which the atom consists. This difference between the mass of an atom and the sum of the masses of its parts is called the *mass defect* as defined by physicists [7].

For example, ^{12}C consists of 6 protons (accurate mass of $1.007277\,u$ each), 6 electrons (accurate mass of $0.000548597\,u$ each), and 6 neutrons (accurate mass of $1.008665\,u$ each); their combined calculated accurate mass is $12.098943\,u$. However, ^{12}C has a defined mass of $12.00000\,u$ [8]. The reason ^{12}C has an accurate mass less than the sum of its protons, neutrons, and electrons is that the binding energy of the nucleus consumes mass (in the form of energy, $E = mc^2$) to hold the positively charged protons together and form the nucleus of the carbon atom. The force responsible for binding of the nucleus of the atom (called the binding energy) is the residual strong force, which creates the mass defect [7].

The mass defect, as this loss of mass is called, can be measured relative to $^{12}C = 12.0000\,u$ by accurate mass spectrometry and is an important component in measuring the accurate mass of molecules and the determination of their molecular

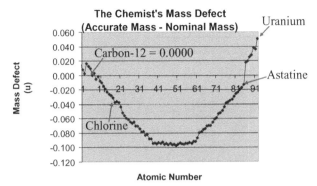

Figure 2.1.   The mass defect is relative to $^{12}C = 12,0000$.

formulas. In 1962, an agreement was reached among physicists and chemists [9] to change the standard accurate masses from an oxygen (16.0000 u) based system to a carbon based system as the standard mass (12.0000 u). This agreement defines the hydrogen atom then as 1.0078 u, which satisfied physicists, since now the mass of hydrogen atom was nearly equal to the resting mass of the proton and electron. Thus, as the nuclei of the periodic table are built with the addition of either a neutron (deuterium) or a proton and electron and two neutrons (helium), the accurate mass of the element reflects the correct mass defect and the binding energy of the nucleus [9].

2.2.2   The Chemist's Mass Defect

Chemists use a different definition of the *mass defect* and define it as the difference between the nominal mass of an element and its accurate mass based on ^{12}C as 12.0000 [10]. The nominal mass of an atom is the mass displayed as the number of protons and neutrons in the nucleus of the atom. For example, H has a nominal mass of 1 and contains one proton. Carbon is 12 and contains 6 protons and 6 neutrons. Nitrogen is 14, O is 16, and so forth for the elements of the periodic table. The nominal mass is useful in visualizing the relative mass defect of the common elements that make up most organic molecules. For example, Figure 2.1 shows the difference between the accurate masses of the common elements minus the nominal mass of the elements, which is the chemist's mass defect.

The chemist's mass defect increases to −80 mmu when going from ^{12}C to ^{79}Br (Figure 2.1). Note that the elements of hydrogen and nitrogen give a positive mass defect and that oxygen, sulfur, chlorine, and bromine give a negative mass defect, while carbon has obviously no effect on the mass defect of the molecule (Figure 2.1).

2.2.3   The Isotopic Mass Defect

Table 2.1 shows the natural abundance of the important isotopes of the common elements found in many organic molecules. Carbon has ^{13}C at 1.078%, nitrogen has

TABLE 2.1. Common Stable Isotopes and Their Natural Abundance

Isotope	Relative Atomic Mass u	Isotopic Composition %
^1H	1.0078250321	99.9885
^2H	2.0141017780	0.0115
^{12}C	12.000000000	98.93
^{13}C	13.0033548378	1.07
^{14}N	14.0030740052	99.632
^{15}N	15.0001088984	0.368
^{16}O	15.9949146221	99.757
^{17}O	16.99913150	0.038
^{18}O	17.9991604	0.205
^{19}F	18.99840320	100
^{23}Na	22.98976967	100
^{28}Si	27.9769265327	92.2297
^{29}Si	28.97649472	4.6832
^{30}Si	29.97377022	3.0872
^{31}P	30.97376151	100
^{32}S	31.97207069	94.93
^{33}S	32.97145850	0.76
^{34}S	33.96786683	4.29
^{35}Cl	34.9688527	75.78
^{37}Cl	36.96590260	24.22
^{79}Br	78.9183376	50.69
^{81}Br	80.916291	49.31
^{127}I	126.904468	100

^{15}N at 0.369%, ^{17}O at 0.038%, ^{33}S at 0.762%, and hydrogen has ^2H (deuterium) 0.0116% (values from NIST page of isotopes, ref. 8). These isotopes make up the important A + 1 isotopes of the common elements found in molecules that are typically studied by mass spectrometry. The elemental calculator of the mass spectrometer takes these abundance values (i.e., 1.078% for ^{13}C) into consideration when calculating the relative abundance ratios of the elemental formula, the A + 1 for example. It is important to note that not only does a molecule have an accurate mass associated with its formula of the most abundant isotope (referred to here as the A peak, a McLafferty convention [11]), but also there is an accurate mass associated with the major isotopes clusters of the molecule. These isotopic masses are important to use in conjunction with the elemental-formula calculator to further confirm the correct molecular formula. However, the elemental-formula calculators of the instrument typically do not take into account the fact that the A + 1 and A + 2 peaks are in fact a summation of all of the isotopes of a molecule, which make various different contributions to the accurate mass of the peak. Only the A peak contains the monoisotopic accurate mass. Thus, it should be possible to calculate the contribution to the accurate mass of each of the important isotopes of a molecule, such as ^{13}C, ^{15}N, etc. to the A + 1 peak.

With these considerations in mind, we define the relative isotopic *mass defect* (RΔm) as the difference in accurate mass between the isotopes of an element minus the nominal mass difference, where the reference isotope is the monoisotopic mass. For example, ^{13}C has a relative positive mass defect as calculated below:

$$\left(^{13}C - {}^{12}C\right) - 1 = R\Delta m_{(A+1)13C}$$

$$(13.00335 - 12.0000) - 1 = R\Delta m_{(A+1)13C}$$

$$1.00335 - 1 = R\Delta m_{(A+1)13C}$$

$$0.00335 = R\Delta m_{(A+1)13C}$$

where RΔm$_{(A+1)13C}$ is the relative isotopic mass defect of the ^{13}C isotope for the A + 1 peak, which is relative to the A peak containing the monoisotopic mass. The A + 2 relative isotopic mass defect when there are two ^{13}C atoms in a molecule is simply twice the amount of the A + 1 defect since there are two ^{13}C atoms present in the molecule, which is a positive 0.0067 mass units.

However, the relative mass defect of an A + 2 peak contributed by a ^{37}Cl is a negative mass defect because of the larger binding energy of the chlorine nucleus relative to its monoisotopic mass of ^{35}Cl (Figure 2.1) and the larger mass defect of the ^{37}Cl, as shown in the calculation below.

$$\left(^{37}Cl - {}^{35}Cl\right) - 2 = R\Delta m_{(A+2)37Cl}$$

$$(36.9659 - 34.9689) - 2 = R\Delta m_{(A+2)37Cl}$$

$$1.9970 - 2 = R\Delta m_{(A+2)37Cl}$$

$$-0.0030 = R\Delta m_{(A+2)37Cl}$$

Figure 2.2 shows the relative isotopic mass defects of each of the common elements and their stable isotopes, which are found in most organic compounds. Note that there are both positive relative isotopic mass defects, ^{13}C, ^{2}H, and ^{17}O but also negative relative isotopic mass defects such as ^{15}N, ^{33}S, ^{37}Cl, and ^{81}Br. Thus, the relative isotopic mass defect is important in the contribution to the A + 1 and A + 2 accurate masses. In conclusion of this section, the name, relative isotopic mass

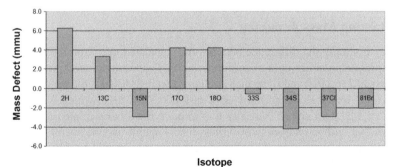

Figure 2.2. Relative isotopic mass defect (accurate mass of the monoisotopic mass minus the accurate mass of the A + 1 or A + 2 isotope).

defect, is consistent with the chemist's definition of the mass defect, and it is called a relative isotopic mass defect because it is relative to the monoisotopic mass of the element.

2.3   EXPERIMENTAL METHODS

2.3.1   Chemicals and Reagents

Pesticide analytical standards were purchased from Dr. Ehrenstorfer (Ausburg, Germany). Individual pesticide stock solutions (200–500 µg/mL) were prepared in pure methanol and stored at −18°C. HPLC grade acetonitrile and methanol were obtained from Merck (Darmstadt, Germany). Formic acid was obtained from Fluka (Buchs, Switzerland). A Milli-Q-Plus ultra-pure water system from Millipore (Milford, MA) was used throughout the study to obtain the HPLC-grade water used during the analyses.

2.3.2   LC/TOF-MS

The LC was a model HP 1100 (Agilent Technologies, Santa Clara, CA), injection volume was 50 µL, the column was a ZORBAX Eclipse® XDB 4.6 × 150 mm C-8 (Agilent), 5-micron with a mobile phase of A = ACN and B = 0.1% formic acid in water. The gradient was 15% A to 100% A over 30 minutes at a flow rate of 0.6 mL/min. The LC/TOF-MS was a model MSD TOF (Agilent Corp, Santa Clara, CA) with an electrospray source positive and negative (ESI+/−), capillary 4000 V, nebulizer 40 psig, drying gas 9 L/min, source temperature 300°C. The fragmentor was used at two voltages of 190 V and 230 V (low and high fragmentor), skimmer 60 V, Oct DC1 37.5V, OCT RF V 250 V. Reference masses (Agilent Solution) were 121.0509 (purine) and 922.0098 (hexakis(1H, 1H, 3H-tetrafluoropropoxy)phosphazine) m/z, with a resolution of 9500 ± 500 at 922.0098 m/z. The reference A sprayer 2 operates at a constant flow rate (600 µL/hr).

2.4   RESULTS AND DISCUSSION

2.4.1   Obtaining Mass Accuracy

Mass accuracy is simply the difference between the calculated mass and the measured mass expressed in either ppm or millimass units. For example, atrazine has a calculated mass of 216.1010 and if the measured mass were 216.1012, the difference would be +0.0002 for a calculated mass error of 0.0002/216.1010, which is equal to $+1.1 \times 10^6$ or +1.1 ppm error. Typically the accurate mass is calculated to four decimal places, but with mass accuracies of less than 1 ppm it may be useful to use five or six place mass accuracies. The atrazine example then would be 216.1012–216.101013 or 0.000213/216.101013, which is still 1.1 ppm error.

When the source is clean, there is good chromatography, clean solvents and blanks, and good detector response, it is time to consider three factors that affect mass accuracy. They are tuning of the mass axis, fine tuning and calibration using a test solution for mass accuracy, such as the NSAID solution [2], and lastly measuring mass in the "sweet spot" of mass accuracy. Combining these three factors makes it possible to measure mass accuracy within ±0.2 millidaltons, which has resulted in the ability to measure the mass of an electron by LC/TOF-MS [8].

Step one is the tuning of the instrument and is done by software of the instrument. Likewise the calibration of the instrument is also done by instrument software. However, after calibration it is useful to use an NSAID solution consisting of three compounds, ibuprofen, diclofenac, and naproxen. These three compounds can be averaged to determine the quality of the calibration. If the values do not average ±0.2 millidaltons, then the calibration is repeated until the instrument is capable of mass accuracy of ±0.2 mDa. These compounds work well in both positive and negative ion electrospray; thus, they make a good calibration mixture.

2.4.2 Measuring the Relative Isotopic Mass Defect

Figure 2.3 shows the mass spectrum of the pesticide spinosad D by LC/TOF-MS positive ion electrospray and the A, A + 1, A + 2, and A + 3 isotope peaks, which contain C, H, O, and one N. The measured mass of the A peak was m/z 746.4831, the A + 1 peak was m/z 747.4862, the A + 2 peak was m/z 748.4893, and the A + 3 peak was m/z 749.4919 for this positively charged compound. The relative isotopic mass defect of the A + 1 was +0.0031 (+3.1 mmu), the relative isotopic mass defect of the A + 2 was +0.0062 (+6.2 mmu), and the relative isotopic mass defect of the A + 3 was +0.0088 (+8.8 mmu).

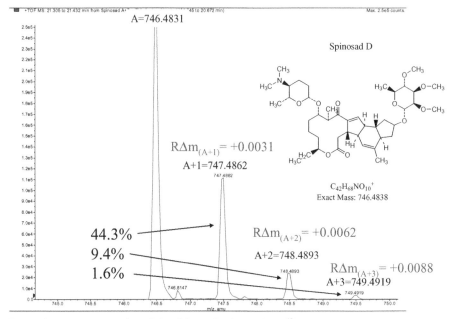

Figure 2.3.   A + 1, A + 2, and A + 3 isotopic signals from ^{13}C.

The molecular formula of this pesticide is $C_{42}H_{67}NO_{10}$ and its protonated molecule has an exact mass of m/z 746.4838. Thus, the measured mass is within 1 ppm of the correct exact mass. As long as the instrument measures accurate masses that are less than 3 ppm then these differences in exact mass are not critical to the calculation of the relative mass defect because of the fact that the isotopic masses are relative to one another at the instrument's limit of accuracy, which is approximately 0.1–0.2 mmu for today's modern TOF instruments [12]. This fact is critically important and herein lies the usefulness of the relative mass defect. This fact will be clear in the following spinosad D example where the intensity of each isotope is included with its relative mass defect.

Because there are 42 carbon atoms in the molecule and the percentage of ^{13}C is 1.078%, then there is a peak at A + 1 mass units with an intensity that is approximately ~45% of the A peak ($42 \times 1.078 = 45.3\%$). The ^{13}C accurate mass is 13.00335 (Table 2.1), which means that the majority of the signal that arises at m/z 747.4862 (A + 1 accurate mass in Figure 2.3) is due to the replacement of one atom of ^{12}C with one atom of ^{13}C. There is also a small contribution to the A + 1 peak from ^{2}H and from ^{17}O. However the ^{17}O isotope is 0.038 percent and only 10 atoms are present; thus, the contribution is small ($10 \times 0.038 = 0.38\%$ of the ~45% total). Likewise, the deuterium contribution is approximately 0.0116% times 68 atoms or 0.79% of the ~45% peak (A + 1), so it too makes a small contribution to the A + 1 peak, as does the ^{15}N (0.37%).

Thus, in this example, the ^{13}C isotope is making the major contribution to accurate mass of the A + 1 peak, which is calculated as follows:

$$P_{13C} = 45.3/(45.3 + 0.38 + 0.79 + 0.37)$$
$$P_{13C} = 96.7\%$$

Where P_{13C} is the abundance in percent of ^{13}C in the A + 1 peak. The total contributions of all four A + 1 isotopes are made by similar calculation (^{13}C, ^{17}O, ^{15}N, and ^{2}H). The $R\Delta m_{(A+1)}$ for the other two isotopes (^{2}H and ^{17}O) are calculated in the same way as the ^{13}C and summed together for the total value of the A + 1 mass increase. The summed relative isotopic mass defects ($\Sigma R\Delta m_{(A+1)}$) for the all of the isotopes in the molecule that contribute to the A + 1 peak is simple to calculate using the data in Table 2.1 and the equation below. We call this summed relative isotope mass defect ($\Sigma R\Delta m_{(A+1)}$), the isotope mass average ($IMA_{(A+1)}$).

$$IMA_{(A+1)} = \Sigma R\Delta m_{(A+1)} = \Sigma (P_{13C}\Delta m_{13C} + P_{2H}\Delta m_{2H} + P_{17O}\Delta m_{17O} + P_{15N}\Delta m_{15N})$$

Thus, for our example in Figure 2.3 of spinosad D, the A + 1 peak has a summed $R\Delta m_{(A+1)}$ of 0.0031 u measured and 0.0033 u calculated, A + 2 is 0.0062 measured and 0.0065 calculated, and A + 3 is 0.0088 measured and 0.0095 calculated. We have designed an Excel spreadsheet that makes the IMA calculation using the summed relative isotopic mass defects and the values of Table 2.1. The IMA calculation works well as long as the resolving power of the instrument is below 100,000, which all LC/TOF and Q/TOF instruments typically are [13]. At higher resolving powers the A + 1 peak begins to separate depending on the mass of the molecule and its isotopes. Thus, our convention is useful for instruments operating in a midrange of resolving power [13].

2.4.3   Examples of Negative Isotope Mass Average (IMA)

Figure 2.4 shows the mass spectrum of propazine with a measured m/z of 230.1164 versus an exact mass of 230.1167 (mass accuracy of 1.3 ppm). Propazine is a triazine herbicide that contains five nitrogen atoms and a single chlorine atom. The ^{15}N atoms and the ^{37}Cl atom create a significant shift not only in the intensity (peak height) of the A + 1 and A + 2 isotope peaks, but also in the $R\Delta m_{(A+1)}$ and $R\Delta m_{(A+2)}$ mass defects. Because there are five nitrogens in propazine and ^{15}N is 0.368%, the intensity of the A + 1 peak is only 11.7% and the measured relative isotopic mass defect is 2.4 mmu. The calculated $IMA_{(A+1)}$ value is also 2.4 mmu, when all the A + 1 atoms (^{13}C, ^{2}H, and ^{15}N) are averaged. Note that the $IMA_{(A+1)}$ value is considerably less than the spinosad D example of 3.1 mmu. This example shows that molecules that contain a significant N/C ratio will deplete the $R\Delta m_{(A+1)}$ value significantly below +3 mmu. Thus, $IMA_{(A+1)}$ values less than 2.5 mmu can be used as a guide to ions with high N content relative to carbon.

Because propazine contains only 9 carbon atoms and 16 hydrogen atoms, neither C nor H make a significant contribution to the A + 2 isotope (see data in Table 2.1). Thus, the relative isotopic mass defect of the A + 2 is due mainly to chlorine (relative peak height of the ^{37}Cl is 32.4% and two atoms of ^{13}C or ^{2}H at the

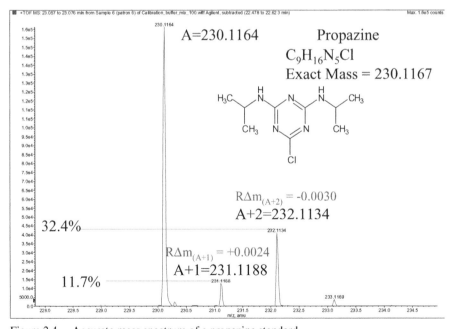

Figure 2.4.   Accurate mass spectrum of a propazine standard.

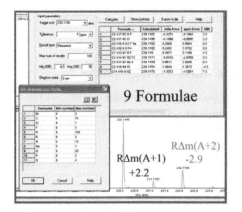
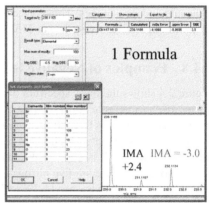

Figure 2.5.   Using the relative mass defect and the isotope mass average (IMA) to limit the number of possible formulae at a 5-ppm error limit by forcing 1 chlorine atom in the molecular formula. Mass spectrum is from a sample extract of Kansas River that contains the herbicide propazine.

A + 2 is about 0.1%). The measured $R\Delta m_{(A+2)}$ versus the calculated $R\Delta m_{(A+2)}$ is −3.0 mmu versus −3.0 mmu. The intensity of the A + 2 peak is also close to the calculated value of 32.4%. Thus, the $IMA_{(A+2)}$ value and the intensity of the isotope peak gives us two important types of information to help in the confirmation of a molecular formula from the A peak. It is clear that this compound contains only one chlorine atom.

The usefulness of $R\Delta m$ and IMA are shown in Figure 2.5, which is the mass spectrum of an extract of the Kansas River that contains a suspect detection of propazine. The accurate mass is m/z 230.1165, which is within 1 ppm of the correct accurate mass. The number of formulae though that match this accurate mass using a 5-ppm mass window and the elements that are commonly found in most pesticides (C, H, N, O, Cl, S, P, and Na for adduct formation) give 9 possible formulae. The relative mass defect of the A + 1 and A + 2 isotopes show values of +2.2 and −2.9, respectively. When the formula generator is forced to contain one chlorine atom the number of formula is reduced to only one formula and its IMA values are +2.4 and −3.0 respectively (Figure 2.5). Thus, this evidence is convincing that the correct identification has been made, at least with respect to molecular formula (i.e., there still exists the possibility of isomers so that a standard and retention time is also required). Thus, the $R\Delta m$ and the IMA tools are useful for empirical formula identification when used with the intensity data of the A + 1 and A + 2 isotopes.

The next example is shown in Figure 2.6 and is the negative-ion mass spectrum of a singly brominated herbicide, bromacil, which was also detected in the extract of the Kansas River water sample. The measured accurate mass is m/z 259.0090 versus the calculated exact mass of 259.0088, an error of 1 ppm. The characteristic mass spectral pattern is shown for the single bromine isotope (approximately equal peak intensity at a nominal mass difference of 2 units) with a relative isotopic mass defect that is considerably less than chlorine at −2 mmu (See Table 2.1 and Figure

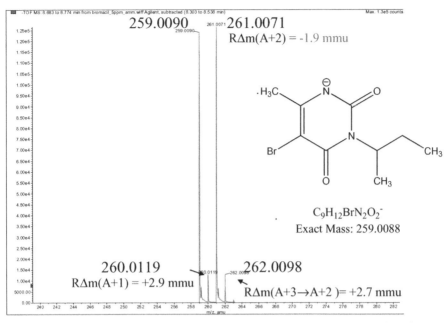

Figure 2.6.   Negative-ion mass spectrum of bromacil, showing relative mass defects for the A + 1, A + 2, and A + 3 isotope cluster.

2.2 for bromine mass defect). The $IMA_{(A+2)}$ value for bromacil is -1.9 mmu. Thus, the measured relative isotopic mass defect is within 0.1 mmu. Note that the $R\Delta m_{(A+1)}$ is +2.9 mmu, which is quite close to the IMA value of +3.0 mmu for bromacil. This example further shows the usefulness of combining the relative intensity of the isotopic pattern with the relative isotopic mass defect for correct identification of halogenated compounds. Also note that there is a RΔm for the A + 3 peak relative to the A + 2 peak, which is written as $R\Delta m_{(A+2\rightarrow A+3)}$, that is +2.7 mmu, which is also quite close to the $IMA_{(A+2\rightarrow A+3)}$ value of +3.0 mmu. Thus, a useful pattern arises that the carbon isotope for any isotopic mass peak is about a positive +3 mmu greater. This is an important heuristic tool that is further elaborated on in the following section on the "Rule of Three."

Figure 2.7 shows the mass spectrum of diuron in olive oil. The measured accurate mass of diuron is m/z 233. 0251 and the calculated exact mass is 233.0243, which is an error of 3.5 ppm. In this example, the characteristic intensity of the A + 2 and A + 4 isotopes are shown in the mass spectra with 66% and 11%, respectively, for the intensities. Likewise there is a negative relative isotopic mass defect of -2.9 mmu for the A + 2 ($IMA_{(A+2)} = -2.9$) and a -6.0 mmu relative isotopic mass defect for the A + 4 isotopes ($IMA_{(A+4)} = -5.8$). Note, however, that the A + 1 relative isotopic mass defect is +2.9 mmu ($IMA_{(A+1)} = +2.9$) and the A + 3 is also a +3.0 mmu relative to the A + 2 ($IMA_{(A+2\rightarrow A+3)} = +2.9$), since this isotopic peak contains a ^{37}Cl and ^{13}C. Thus, the isotopic pattern coupled with the relative isotopic mass defect and the calculated IMA value is useful and important for identification of pesticides in food and water even when the accurate mass is slightly greater than

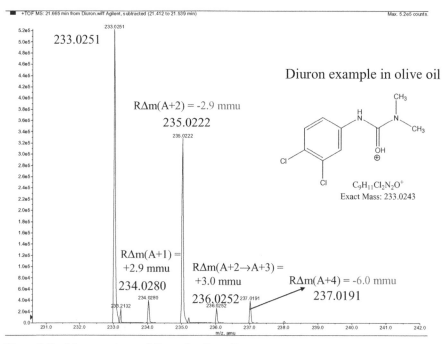

Figure 2.7.   Mass spectrum of diuron in olive-oil extract.

3 ppm. In spite of this mass inaccuracy, the IMA values are within 0.1–0.2 mmu and are valuable for correct formula generation. It is clear that two chlorines can be specified in this example with the result of easily generating the correct formula for diuron even at an error of 3.5 ppm mass accuracy, which is an error that is not uncommon for difficult food extracts such as olive oil.

2.4.4   "Rule of Three"

It is useful to introduce a simplification to the calculation of IMA and the relative isotopic mass defect for rapidly looking over the accurate masses of peaks in the mass spectrum. The simplification comes from the statistical analysis of over 100 accurate mass measurements of the A + 1 and A + 2 isotopes of pesticides in solvent, water, and food matrices. The result is that for A + 2 chlorinated compounds the average for the relative mass defect is −2.9 mmu ±0.1 mmu (standard deviation). The result for the A + 1 for these same compounds is a relative mass defect of +3.1 mmu ±0.2 mmu. Brominated A + 2 values will be less, around −2 mmu, and mixed brominated and chlorinated compounds will vary from −2 to −3 mmu. The A + 1 values for these compounds remain in the +3.0 mmu range.

Thus, the "Rule of Thumb" or heuristic rule that is useful for the relative isotopic mass defect is the +3 mmu for A + 1 and A + 3 isotopes (relative to the A and A + 2, respectively) and −3 mmu for A + 2 and A + 4 isotopes (relative to the A and the A + 2, respectively). The idea of the heuristic rule is that one examines

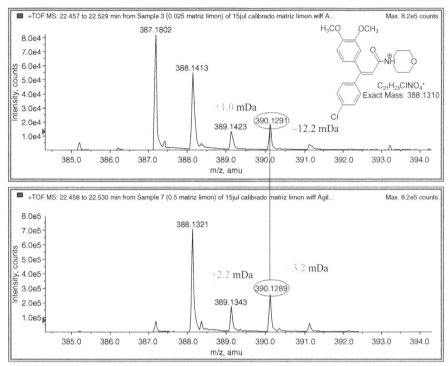

Figure 2.8.   Interference from the C-13 isotope of the *m/z* 387 ion for dimethomorph, *m/z* 388.

the raw accurate mass data of a mass spectrum "by eye," which is quite easy to do. Then, one only looks at the mass defect of the last two decimal numbers of the accurate mass at four decimal places and determines if the mass defect is positive or negative and by how much. Not only can the rule be used for formula generator software to eliminate incorrect formula, it can also be used to spot mass interferences in the accurate mass data and make corrections, as shown in the following section.

2.4.5   Using IMA for Mass Interferences

The last example, shown in Figure 2.8, is an example where an interfering ion is present in the mass spectrum of the pesticide and how to use the "Rule of Three" (the relative isotopic mass defect) to point out and correct the problem. The upper section in Figure 2.8 shows a mass spectrum of the pesticide, dimethomorph, in a lemon extract *m/z* 388.1413 and a possible interfering ion at *m/z* 387.1802, based on the incorrect relative mass defects for dimethomorph of +1.0 and −12.2 for the A + 1 and A + 2 ions, respectively. The relative mass defects are clearly wrong, especially the A + 2 value. This interfering ion at *m/z* 387 must have a ^{13}C peak that is the major part of the spectrum for dimethomorph at *m/z* 388.1413. The calculated exact mass for dimethomorph is *m/z* 388.1310. The calculated ^{13}C ion for the *m/z* 387 interfering ion is 388.1835, which is only ~25 mmu different than the dimethomorph at 388.1310. Thus, there is a serious mass interference from the m/z 387 ion. In fact, the correct formula for dimethomorph was listed as the 25th choice with an error of 26 ppm. Thus, the value of the A + 1 and A + 2 isotope data shows that a serious interference is present in the A peak.

When the value for the lower part of Figure 2.8 was chosen, m/z 388.1321, for dimethomorph then the correct formula was ranked at number four with 2.8-ppm error. The bottom mass spectrum was the dimethomorph spiked into the lemon extract at a higher concentration of 0.5 ppm, which was sufficiently high enough to reduce the interfering ion and to give relative mass defect values that were closer to the correct value. The observed values for the relative mass defect were +2.2 and −3.2 (Figure 2.8), versus the IMA value of +3.3 and −2.3 (still showing mass interference but less). Note that in both the upper and lower spectra the value of the A + 2 peak is approximately the same number of m/z 390.1291 and 390.1288. Thus, this result suggests that one could use the A + 2 values for formula generation in cases of interference from an A + 1 interfering ion.

Another cause for the relative isotopic mass defect to be incorrect is a high concentration and subsequent overloading of the A peak when measuring the accurate mass. Although today's TOF instruments are capable of three orders of magnitude of linear concentration range [14], there can still be an overloading of compound, which skews the mass to a higher and incorrect mass. This error causes the relative mass defect at A + 1 to be much smaller value than the rule of +3 mmu. Because the A + 1 peak is usually 10 times smaller than the A peak, the overloading often will be absent in the A + 1 peak. The error in the relative mass defect will be considerably greater than the 0.2–0.3 mmu errors that are typically seen. When this is observed in a mass spectrum, it is important either to shift to a different part of the A mass peak with less intensity or to dilute and re-analyze the sample. Thus, in conclusion, the use of the relative mass defect and the IMA value is a useful procedure to check the molecular formula for interferences and accuracy, and to limit the number of atoms in the elemental calculator for the correct formula from exact mass data.

2.5 CONCLUSIONS

The relative isotopic mass defect is either positive by approximately 3 mmu units for the A + 1 and A + 3 isotopes or negative from −2 to −3 mmu for the A + 2 elements, such as chlorine, bromine, and sulfur. The examples in this paper show that with the modern LC/TOF-MS instruments of today, with accuracies of less than 3 ppm possible [12], the IMA and relative isotopic mass defect are useful tools for compound identification because the number of elemental possibilities can be markedly decreased by forcing of the empirical formula. Furthermore, the IMA can be used to find losses of mass accuracy from interferences or conversely to know that the isotopic signal is clean and free of interferences from other masses. These two tools, then, are quite useful and show that instruments with resolving power of 10,000 and mass accuracies of less than 3 ppm are valuable instruments for unknown analysis.

REFERENCES

1. Thurman, E.M.; Ferrer, I.; Fernandez-Alba, A.R., Matching unknown formulas to chemical structure using LC/MS TOF accurate mass and database searching: example of unknown pesticides on tomato skins, *J. Chromatogr. A*, **2005**, 1067, 127–134.

2. Ferrer, I.; Fernandez-Alba, A.R.; Zweigenbaum, J.A.; Thurman, E.M., Exact mass library for pesticides using a molecular feature database, *Rapid Commun. Mass Spectrom.*, **2006**, 20, 3659–3668.

3. Ferrer, I.; Heine, C.E.; Thurman, E.M., Combination of LC/TOFMS and LC/ion trap MS/MS for the identification of diphenhydramine in sediment samples, *Anal. Chem.*, **2004**, 76, 1437–1444.

4. Thurman, E.M., Accurate mass determination of chlorinated and brominated products of 4-nonylphenol, nonylphenol dimers, and other endocrine disruptors, *J. Mass Spectrom.*, **2006**, 41, 1287–1297.

5. Ferrer, I.; Thurman, E.M., *Liquid Chromatography/Mass Spectrometry for the Analysis of Emerging Contaminants, American Chemical Society Symposium Volume 850*, Oxford University Press, **2003**, Washington, DC.

6. Ferrer, I.; Thurman, E.M., Liquid chromatography/time-of-flight mass spectrometry (LC/TOF/MS) for the analysis of emerging contaminants, *Trends Anal. Chem.*, **2003**, 22, 750–756.

7. Physics Basics, Structure of the Atom, Nuclear Physics, web site: http://www.tpub.com/content/doe/h1019v1/css/h1019v1_42.htm

8. NIST web site of isotope masses and percentages: http://physics.nist.gov/cgi-bin/Compositions/stand_alone.pl?ele=&all=all&ascii=html&isotype=some.

9. Cameron, A.E.; Wichers, E., Report of the International Commission on Atomic Weights, *J. Am. Chem. Soc.*, **1962**, 84, 4175–4176.

10. Sparkman, O.D., *Mass Spectrometry Desk Reference, 2002*, Global Publishing, Pittsburgh, PA.

11. McLafferty, F.W., *Interpretation of Mass Spectra*, Third Edition, **1980**, University Science Books, Mill Valley, CA, 303 p.

12. Ferrer, I.; Thurman, E.M., Measuring the mass of an electron by LC/TOFMS: A study of "twin ions", *Anal. Chem.*, **2005**, 77, 3394–3400.

13. Thurman, E.M.; Ferrer, I.; Zweigenbaum, J.A., High resolution and accurate mass analysis of xenobiotics in food, *Anal. Chem.*, **2006**, 78: 6703–6708.

14. Ferrer, I.; Thurman, E.M.; Fernandez-Alba, A., Quantitation and accurate mass analysis of pesticides in vegetables by LC/TOF-MS, *Anal. Chem.*, **2005**, 77, 2818–2825.

PART II

TOOLS FOR UNKNOWN IDENTIFICATION USING ACCURATE MASS

CHAPTER 3

ION COMPOSITIONS DETERMINED WITH INCREASING SIMPLICITY

Andrew H. Grange and G. Wayne Sovocool

U.S. EPA, ORD, NERL, Environmental Sciences Division, Las Vegas, Nevada

EXACT MASSES of ions and relative isotopic abundances (RIAs) of the +1 and +2 isotopic mass peaks, measured with sufficient accuracy, provide the elemental compositions of ions and limit the number of possible compound identities. Double focusing mass spectrometers provided both exact mass and RIA measurements in the early 1990s, while today, less expensive, more robust, and easier to operate time-of-flight mass spectrometers (TOFMSs) provide data sufficiently accurate to determine ion compositions. Using exact masses and RIAs measured with a TOFMS, an Ion Correlation Program matches precursor ion and product ion:neutral loss pairs to increase the mass range for which the unique and correct composition of ions can be determined and to provide deconvolution of composite mass spectra. This chapter briefly reviews the evolution in the types of mass spectrometers, data acquisitions, and automated determination of ion compositions used in our laboratory to more easily and rapidly identify compounds.

3.1 INTRODUCTION

For 15 years our laboratory has identified compounds in environmental extracts with little or no knowledge of the sampling site history [1–12]. Elements found in molecules have included C, H, N, O, Cl, Br, F, I, S, P, Si, As, and Se. Data acquired with a double focusing, high-resolution mass spectrometer using a successive approximation approach provided exact masses and relative isotopic abundances (RIAs) of ions of sufficient accuracy to establish the elemental composition of the molecular ion or highest-mass product ion. In-house software automated set up for analyte-specific data acquisition and for data processing, including comparison of measured and calculated exact masses for often numerous possible compositions.

Liquid Chromatography Time-of-Flight Mass Spectrometry: Principles, Tools, and Applications for Accurate Mass Analysis, Edited by Imma Ferrer and E. Michael Thurman
Copyright © 2009 John Wiley & Sons, Inc.

Over the past few years, the instrumentation, data acquisition, and automated data evaluation necessary to determine ion compositions have evolved toward greater simplicity. Consequently, others are now identifying contaminants found in water [13] and food [14, 15] with relative ease. This chapter describes our parallel progressions in instrumentation, data acquisition methods, and data interpretation. Double focusing mass spectrometers, an accurate-mass triple quadrupole mass spectrometer, and an orthogonal-acceleration, time-of-flight mass spectrometer have been used to acquire accurate-mass data using selected ion recording, multiple reaction monitoring, and full scanning, respectively. Data processing has evolved from using a mass-resolution-dependent Profile Generation Model to determine ion compositions to using a simpler Ion Correlation Program to determine precursor ion, product ion, and corresponding neutral loss compositions for mass spectra produced by single or multiple analytes.

3.2   COMPOUND IDENTIFICATION

The NIST [16] or Wiley [17] libraries often provide mass spectra similar or related to that of an analyte, which can lead to a tentative identification, even when only nominal (or integer) masses are measured. If not, determining the elemental composition of a molecule that has lost an electron or gained a proton to become a positively charged ion, or lost a proton or captured an electron to become a negatively charged ion, limits a compound's identity to a finite number of isomers. Product ions formed from such a precursor ion and the corresponding neutral losses (product ion:neutral loss pairs) provide insight into the structural features of the compound, thereby limiting the number of possible isomers. Consultation of databases [18, 19] such as the SciFinder® compilation of known isomers [20] further reduces the number of plausible isomers. A tentative identification is confirmed when a purchased or synthesized standard with a verified structure provides the same mass spectrum and retention time by gas chromatography or liquid chromatography as the analyte.

3.3   EXACT MASSES AND RELATIVE ISOTOPIC ABUNDANCES

The exact mass is only one of two measurable and independent physical properties of an ion that provide differentiation among possible ion compositions. In addition to the exact mass of the monoisotopic ion (M), which contains only atoms of the lowest-mass isotopes, the Relative Isotopic Abundances (RIAs) of the M + 1 and M + 2 mass peaks that arise from the presence within an ion of atom(s) of isotopes heavier by 1 or 2 Da than the lowest-mass isotopes are unique for different sets of atoms. The RIAs of the +1 and +2 isotopes of the elements [21] most commonly found in organic molecules, C, H, N, O, Br, Cl, F, I, P, S, and Si result in the abundance of the M + 1 mass peak providing an estimate of the number of C and Si atoms in an ion, while the abundance of the M + 2 mass peak provides estimates of

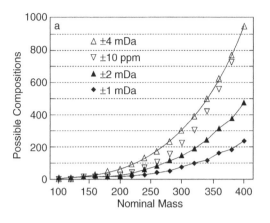
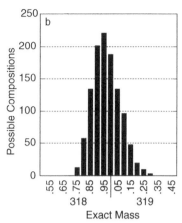

Figure 3.1. (a) the number of possible compositions as a function of mass for four different error limits. The elements C, H, N, O, P, S, and Cl were considered for masses between 100.0000 and 400.0000 Da at 20-Da intervals. At m/z 400, 4 mDa equals 10 ppm and the number of possible compositions is the same for the two error limits. For masses below 400 Da, 4 mDa is greater than 10 ppm and fewer compositions are possible for the 10 ppm error limit. The opposite is true from masses greater than 400 Da. (b) The number of possible compositions as a function of the mass defect calculated at 0.05-Da intervals between 318.5500 and 319.4500 Da for the same elements and a mass error limit of ±2 mDa [19].

the number of S, Si, Cl, and Br atoms present. Note that the generic (M) can be a molecule that has lost an electron or gained a proton (positive ion) or lost a proton or captured an electron (negative ion).

The mass error limits and RIA error limits determine the relative utility of exact masses and RIAs for determining the unique and correct ion composition. Figure 3.1 illustrates that the number of possible compositions increases exponentially with increasing mass (Figure 3.1a), increases nearly linearly with increasing mass error (Figure 3.1a), and is highly dependent on the mass defect measured for an ion (Figure 3.1b). Even for a small mass error limit, numerous ion compositions can be possible for high-mass ions. Considering additional elements as possibly present also increases the number of possible compositions, while considering only odd or even electron ions approximately halves the number. The number of possible compositions based on consideration of RIAs alone also increases as an ion's mass or the RIA error limit increases.

Figure 3.2 illustrates for one set of mass error and RIA error limits (±2 mDa and ±20%) a major reduction in the number of possible compositions that occurred when both exact masses and RIAs were considered for measured values of 319.1039 Da, 19.00%, (M + 1), and 36.78% (M + 2) for the [M + H]$^+$ ion from chlorpromazine, an antipsychotic drug [19]. Even assuming that at least one third of the mass of the ion was due to C atoms, 47 compositions were possible for the protonated molecule (82 compositions without the assumption). Considering only the RIAs, 402 compositions were possible. But when both the exact masses and RIAs were

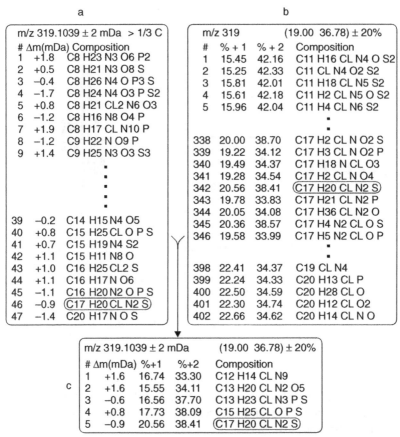

Figure 3.2. (a) partial list of possible compositions for a measured monoisotopic exact mass of 319.1039 Da assuming one third of the ion's mass is due to C atoms and a mass error limit of ±2 mDa with consideration of C, H, N, O, P, S, and Cl atoms. (b) Partial list of possible compositions for a nominal mass of 319 Da and measured RIAs of 19.00% and 36.78% assuming an RIA error limit of ±20% for consideration of the same elements. (c) List of possible compositions found in both partial lists. The correct composition is enclosed by a modified ellipse [19].

considered, only 5 compositions remained viable. Clearly, for the purpose of determining ion compositions, it is advantageous to measure both exact masses and RIAs.

3.4 DOUBLE FOCUSING MASS SPECTROMETER

To accurately measure exact masses and RIAs for coeluting compounds in complex environmental extracts, high mass resolving power was necessary. In the early 1990s, double focusing mass spectrometers alone provided resolving powers of

10,000 to 20,000 (10% valley) for routine analyses. In the environmental arena, these instruments were often used for dioxin analyses [22, 23]. To provide low error limits for quantitation ^{13}C-labeled analogs of dioxins were spiked into extracts. Selected-Ion-Recording (SIR), the monitoring of selected m/z ratios corresponding to the exact masses of the labeled reference and target analyte compounds, provided 100-fold greater sensitivity, fast scan cycles, and 10-fold higher resolving power (10,000, 10% valley) compared to full scanning. We devised methodology to use double focusing mass spectrometers and SIR to provide compositions of ions from unidentified compounds in complex environmental extracts without reference to lists of target compounds [24–26]. With 2200 high production chemicals (10^6 lbs/yr or more) [27] and 87,000 compounds used in commerce [28], most compounds, their synthetic byproducts, and their transformation products do not appear on target lists.

To measure both exact masses and RIAs with high resolving power, four data acquisitions were necessary to provide the plots of full or partial mass peak profiles shown in Figure 3.3. First, a full scan, background-subtracted, mass spectrum using the lowest resolving power (1000) to maximize sensitivity and scan speed was obtained for compounds as they eluted from a gas chromatograph (GC) into the electron ionization ion source. For each compound, the apparent molecular ion (largest-mass, monoisotopic ion observed) was further investigated. Three SIR data acquisitions that covered narrow mass ranges and that delineated full or partial mass peak profiles were then made as analytes eluted from a GC interfaced to either a VG 70-SE mass spectrometer for which 23 of 25 available m/z ratios were used [1–3, 10, 24–26, 29–32] or a Finnigan MAT 950S mass spectrometer for which 31 of 32 available m/z ratios were used [4–9, 11, 12].

In the middle part of Figure 3.3a, a narrow mass range about a monoisotopic ion from an analyte found in an extract of ground water was monitored with a resolving power of 3000 and an m/z increment of 100 ppm between adjacent points. The ion chromatogram for each m/z ratio was integrated across the chromatographic peak for the monoisotopic ion of interest and plotted in Figure 3.3a. The two mass peak profiles, each delineated by three points, were obtained for an analyte ion (left peak inset) and a perfluorokerosene (PFK) ion (right peak inset). The insets are the ion chromatograms for the m/z ratios at the maxima of the two peaks. For the PFK peak, two baseline excursions are evident in the inset that were induced by changing a lens voltage in the ion source to temporarily divert the ion beam. The excursions enabled area integration across the simulated chromatographic peak between them. The areas were plotted to provide the partial mass peak profiles for PFK lock mass (left plot) and calibration (right plot) ions that bracket the analyte ion mass. The exact masses of all full or partial mass peak profiles in Figure 3.3 were obtained as a weighted average of the top several points delineating the profiles.

The exact mass from the analyte mass peak profile in Figure 3.3a was used as the center mass in the SIR m/z ratio list to acquire the ion chromatograms used to plot the analyte ion profile in Figure 3.3b. The resolving power was 10,000 and the mass increment between points was 10 ppm. The exact mass obtained corresponded to eight possible compositions assuming atoms of C, H, N, O, As, F, P, or S could be present and a mass error limit of 6 ppm.

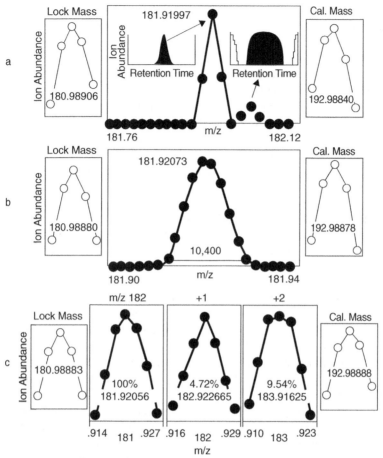

Figure 3.3. Three outputs provided by a Lotus 123® spreadsheet using SIR data: (a) full profile, 100-ppm mass increment, and 3000 resolving power; (b) Full profile, 10-ppm increments, and 10,000 resolving power; and (c) Partial profiles, 10-ppm increment, and 10,000 resolving power.

Experience has shown that the possible composition containing the fewest heteroatoms other than halogen atoms is most likely to be correct when identifying compounds in complex environmental extracts. Hence, it was chosen as the hypothetical composition to create the SIR m/z ratio list to acquire data for the M, M + 1, and M + 2 partial profiles plotted in Figure 3.3c. Summing the chromatographic peak areas used to plot the partial profiles, dividing the summed areas for the M + 1 and M + 2 partial profiles by the summed area of the M partial profile, and multiplying by 100% provided the %M + 1 and %M + 2 relative abundances shown on the partial profiles in Figure 3.3c. This use of SIR to delineate profiles to provide exact masses and RIAs was called Mass Peak Profiling from Selected Ion Recording Data (MPPSIRD) [25].

3.5 APPLYING EXACT MASS AND RIA CRITERIA

To compare measured exact masses and RIAs with calculated values for possible compositions, a Profile Generation Model (PGM) was written in QuickBASIC® version 4.5 (Microsoft Corp., Bellevue, WA) to calculate the mass of the M profile and to construct the M + 1 and M + 2 profiles resulting from the ions containing atoms of higher isotopes [26]. The M + 1 and M + 2 profiles were calculated as sums of the Gaussian distributions for the individual +1 and +2 ions at the resolving power being used. Table 3.1 prepared by the PGM provided a listing of the possible compositions. The "X"s indicated that the measured and calculated values for a composition did not agree within the mass error limit of 6 ppm and the RIA error limit of about 11%. These error limits were established from measurements for several standards [26]. Using MPPSIRD and the PGM together was referred to as Ion Composition Elucidation (ICE) in several papers [6–10].

Initially, two other criteria based on the width and shape of the full +2 profile were examined: the apparent resolution determined from the profile width at 5% of the maximum profile height, which was less than the instrument resolving power when profile broadening occurred, and a peak shape parameter, which was the sum of amplitude differences between measured profiles and profiles calculated for each possible composition [26, 30, 31]. To observe such broadened profiles, usually the M + 2 profile was acquired as a full profile, which required an additional data acquisition. Broad profiles were most often observed when one or more S atoms were present. It became apparent, however, that these two measures seldom, if ever, eliminated compositions that had passed the exact mass and RIA criteria, and they were soon abandoned as a simplification measure. In Table 3.1 and numerous similar tables for other analytes, the exact mass and RIA criteria of the M + 1 and M + 2 profiles eliminated many of the same compositions, but additional compositions

TABLE 3.1. Calculated Exact Masses and RIAs for Eight Compositions* and the Measured Values

Commposition	182 M	183 M + 1	184 M + 2	%M + 1 (%1 Range)#	%M + 2 (%2 Range)#
$HNOF_3P_2S$	.92007	.91896 X‡	.91623	1.21 (1.02–1.39) X	4.63 (4.02–5.25) X
$HNOAsF_4$	.92047	.91842 X	.92472 X	0.42 (0.36–0.48) X	0.20 (0.17–1.23) X
$H_2NO_2F_2S_3$	.92102	.92029 X	.91707	2.83 (2.38–3.29) X	13.72 (11.92–15.53) X
$N_4O_2P_2S$	.92117	.91924 X	.91767 X	2.33 (1.99–2.68) X	4.85 (4.21–5.50) X
N_4O_2AsF	.92157	.91896 X	.92561 X	1.55 (1.33–1.76) X	0.41 (0.34–0.48) X
$C_2NO_5S_2$	.92124	.92265	.91794 X	4.34 (3.69–4.99)	9.94 (8.60–11.28)
C_3HNOFS_3	.91988	.93132 X	.91588	6.09 (5.19–6.99) X	13.65 (11.86–15.44) X
$C_3H_7AsS_2$✓	.91996	.92210	.91585	4.97 (4.23–5.70)	8.96 (7.79–10.13)
Measured values:	.92056	.92265	.91625	4.72	9.54

* Some of the compositions are different from those in the original paper [7], because the mass of the electron, 0.00055 Da, had not been taken into account [33].

The permissible ranges result from consideration of several errors associated with plotting partial profiles [26].

‡ An "X" indicates the measured and calculated values did not agree within mass or RIA error limits.

were eliminated by the RIA criteria, which provided a higher level of discrimination. No chemical reasoning was included in the PGM to eliminate compositions beyond requiring that the rings and double bonds be zero or more for electron ionization or −0.5 or more for ESI or APCI.

The sensitivity and wide dynamic range provided by a double focusing mass spectrometer enabled identification of compounds present over about three orders of magnitude of ion abundance. Figure 3.4 is a partial total ion chromatogram labeled with ion compositions for the apparent molecular ion for most of the chromatographic peaks along with many identifications or tentative identifications [10].

While the three exact mass and two RIA criteria were passed for the correct ion composition in Table 3.1, for low-concentration analytes, passing of only two to four criteria was common. Especially for compounds with %M + 2 values of less than 2%, even a resolving power of 10,000, was insufficient to eliminate interferences from molecular and product ions formed from coeluting compounds, PFK, or column bleed. The RIAs have been found to be more sensitive to interferences than the exact masses [10]. When more than one composition passed the same number of criteria, the presence of related compounds was used as a criterion to select a single composition.

3.6   OBSOLESCENCE OF ICE

Although two other groups utilized MPPSIRD to determine compositions of several molecular ions [34–36], widespread adoption of ICE was discouraged by the required custom software and time needed to acquire data. Four GC/MS data acquisitions were required to determine the exact masses and RIAs of 15–30 ions from individually eluting compounds. To determine ion compositions of apparent molecular ions from 100 chromatographic peaks required several weeks.

Text files of macro language instructions for either data system were prepared by a Lotus 123® spreadsheet (Lotus Development Corp., Cambridge, MA). The data systems then automatically prepared SIR menus, displayed and integrated chromatographic peaks over user-specified time-windows, and provided lists of *m/z* ratios and chromatographic peak areas from which profiles were plotted and exact masses and RIAs were calculated within a second Lotus 123 spreadsheet. Unfortunately, the data systems of current double focusing mass spectrometers no longer provide a macro language. Hence, performing ICE with newer instruments is not practical.

3.7   ACCURATE MASS TRIPLE QUADRUPOLE MASS SPECTROMETER (AM3QMS)

A Thermo-Finnigan TSQ Quantum Ultra AM® triple quadrupole mass spectrometer (Thermo-Finnigan, San Jose, CA) provided masses accurate to within 5–10 mDa and RIAs usually accurate to within 10% (in the absence of interferences) for

Ion Abundance

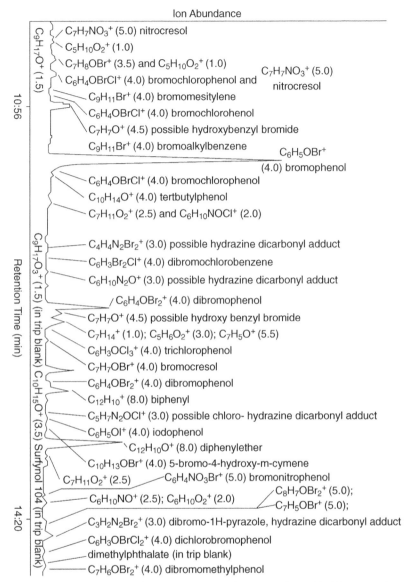

Figure 3.4.   A partial total ion chromatogram for a methylene chloride extract from water above a plume of contaminants from a chemical plant [10].

precursor and product ions from nine standards. A solution of the standards was introduced by HPLC and atmospheric pressure chemical ionization (APCI) was used [19]. Data acquisition was simpler, since no custom software was required, but three types of scans requiring manually prepared *m/z* ratio lists and scan parameters were still needed to obtain exact mass and RIA values. First, a single full scan data acquisition was recorded from the third quadrupole (Q3) using a 0.7 mDa mass peak width as the collisionally induced dissociation (CID) voltage in Q2 was switched among −12, −24, or −36 V. Q2 contained 2 mTorr of argon. Mass spectra containing an abundant precursor ion (−12 V) and product ions (−24 V and −36 V) were obtained.

During the single data acquisition to measure exact masses, internal mass calibration was against ions formed from six compounds in a solution that was infused as individual analyte peaks eluted from the HPLC. Selected Reaction Monitoring (SRM) was used to monitor each analyte precursor or product ion and calibration ions that bracketed the analyte ion mass. The mass peak widths for Q1 and Q3 were 0.7 Da and 0.1 Da, respectively. Q3 was scanned over a 1 Da mass window. Between 6 and 12 different m/z ratios were monitored during elution of each analyte.

For the single data acquisition using SRM that measured RIAs, the mass peak widths were set to 10 Da and 0.5 Da for Q1 and Q3, respectively, and a 1 Da mass range was scanned by Q3. The wide Q1 peak width centered on the M + 1 mass peak allowed all M, M + 1, and M + 2 ions to pass into Q3. Three mass peaks from each of three analytes were monitored during each data acquisition. For both SRM data acquisitions, the CID voltages that produced the greatest abundance for each ion were used when determining their exact mass or RIAs. More detail on these experiments is provided in ref. 18.

3.8 ORTHOGONAL-ACCELERATION, TIME-OF-FLIGHT MASS SPECTROMETER (oa-TOFMS)

Three full scan data acquisitions, one each with different in-source CID voltages, were required to measure exact masses and RIAs when using an Agilent G1969A LC/MSD TOF with an Agilent G1948A electrospray ionization source. After initially selecting a mass range, scan rate, and other parameters to optimize instrument performance, all settings remained constant except for the fragmentor (CID) voltage. Manual or automated entries of precursor ion masses or retention time window boundaries into analyte-specific menus were not required to select ions for fragmentation and mass analysis by a nonexistent second MS stage. No prior knowledge of target ions and retention times for eight standards was required, because after mass calibration, the full scan data provided the exact masses and RIAs for every precursor ion and their product ions. The exact mass errors were 1–2 mDa and the RIAs were accurate to within 20% in the absence of interferences. Both mass error and RIA error limits were sufficient to greatly limit the number of possible compositions for an ion. Of the three types of mass spectrometer, the oa-TOFMS provided a major advantage in simplicity and greatly reduced the time required to measure exact masses and RIAs. More detail on acquiring such data and using it to identify compounds is provided in ref. 19. Note that if GC/oa-TOFMS with electron ionization were used, only one full scan would provide the exact masses and RIAs for the product ions and, in many cases, the molecular ion.

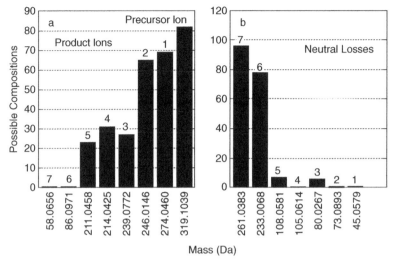

Figure 3.5.   (a) A plot of the possible compositions calculated for the measured exact masses of a precursor ion and seven of its product ions and (b) the seven corresponding neutral losses. The mass error limit is ±2 mDa for the ions and ±2.83 mDa ($\sqrt{2} \times 2$ mDa) for the neutral losses, which are determined as the difference in measured exact masses for the precursor and product ions [19]. The numbers atop each vertical bar correlate the product ion:neutral loss pairs.

3.9   AN ION CORRELATION PROGRAM

The higher error limits associated with both the AM3QMS and the oa-TOFMS necessitated devising a new data processing strategy to determine unique precursor ([M + H]$^+$) ion compositions. Figure 3.5 illustrates that in accord with Figure 3.1a, the numbers of possible compositions for an exact mass within a specified error limit are less for product ions and their corresponding neutral losses than for the precursor ion. When the exact mass and RIAs of a precursor ion correspond to multiple possible compositions, the correct composition can be selected by determining and summing the unique compositions of a product ion:neutral loss pair produced from the precursor ion.

An in-house Ion Correlation Program written in QuickBASIC® 4.5 and based on a simplified PGM rejects numerous compositions, generally leaving only one for the precursor ion, each product ion, and each neutral loss, by comparing their compositions for consistency. All product ion:neutral loss pairs must sum to the composition of the precursor ion. The four-step computational process was described in ref. 19.

Because the resolving power of the AM3QMS was 3000 (FWHM) and that of the oa-TOFMS was 6000 (FWHM), no peak broadening was observed in mass peak profiles or considered in exact mass and RIA calculations. Calculation of Gaussian distributions for a given resolving power was no longer necessary. To provide additional simplification and to conform with common practice [37–39], the exact

TABLE 3.2. Rejection of Incorrect Compositions by M + 1 and M + 2 Exact Mass and RIA Criteria for 15 Compositions[&]

Nominal Mass	Composition	M + 1 and M + 2 Exact Masses # Rejected (# Possible)	M + 1 and M + 2 RIAs # Rejected (# Possible)
124	$C_6H_{10}N_3^+$	2 (4)	3 (4)
166	$C_{10}H_{16}NO^+$	3 (6)	5 (6)
170	$C_{12}H_{12}N^+$	5 (10)	9 (10)
181	$C_{12}H_9N_2^+$	11 (17)	16 (17)
182	$C_8H_8NS_2^+$	30 (38)	37 (38)
195	$C_8H_{11}N_4O_2^+$	10 (19)	18 (19)
214	$C_{10}H_{16}NO_2S^+$	22 (27)	25 (27)[*]
237	$C_{15}H_{13}N_2O^+$	28 (42)	41 (42)
265	$C_{11}H_{13}N_4O_2S^+$	69 (96)	88 (96)[*]
285	$C_6H_{13}Cl_3O_4P^{+\#}$	9 (20)	13 (20)
319	$C_{17}H_{20}N_2ClS^{+\wedge}$	12 (49)	42 (49)
330	$C_{19}H_{21}NO_3F^+$	74 (105)	100 (105)
380	$C_{24}H_{22}N_5^+$	103 (158)	152 (158)
419	$C_{25}H_{39}O_5^+$	43 (80)	74 (80)
436	$C_{24}H_{30}N_5O_3^+$	131 (194)	182 (194)

[&] Elements considered: C, H, N, O, F, P, S, and Cl when present. Mass window: ±2 mDa. RIA window: ±20%.
[*] One composition not rejected by RIA criteria was rejected by exact mass criteria.
[#] Three Cl atoms were assumed to be present as would be apparent from the M + 2 and M + 4 peak abundances.
[∧] One Cl atom was assumed to be present as would be apparent from the M + 2 and M + 4 peak abundances.

masses of the M + 1 and M + 2 profiles were no longer considered. The exact masses of the M + 1 and M + 2 mass peaks are dependent on both the masses of the +1 and +2 isotopes and their isotopic abundances. The M + 1 and M + 2 mass peak dependence on the RIAs resulted in the similarity of composition rejections observed in Table 3.1 for the M + 1 and M + 2 exact masses and RIAs. In addition, mass errors in mDa, rather than ppm, were adopted to avoid very large ppm errors for low-mass product ions. For example, a 5 mDa mass error for m/z 77 would be 65 ppm. The observed mass error between 50 and 500 Da will be less variable when expressed in mDa than in ppm for exact masses measured with a TOFMS.

Table 3.2 compares the discriminating power for rejecting compositions of the M + 1 and M + 2 exact masses (third column) and RIAs (last column) for ±2 mDa and ±20% error limits for 15 compositions containing different sets of heteroatoms. The total number of possible compositions corresponds to a ±2 mDa mass window about the calculated monoisotopic ion masses. For all 15 compositions, comparison of the M + 1 and M + 2 exact masses rejected fewer incorrect compositions than comparison of the RIAs. In eight of nine cases, unique compositions were not found for precursor ions with masses greater than 200 Da. To obtain unique compositions for these precursor ions, product ion:neutral loss pairs must be considered by the ICP. Only for two compositions did considering the M + 1 and M + 2 exact masses reject a composition not already eliminated by the M + 1 and M + 2 RIA criteria. The demonstration that the M + 1 and M + 2 RIAs provide greater discrimination among compositions than the exact masses would be even more compelling with the larger mass error (±10 mDa) and smaller RIA error ($\pm10\%$) for the AM3QMS data. Conversely, if a different type of mass spectrometer provided tighter mass error limits, but was incapable of providing RIAs accurate to within 20%, consideration of the M + 1 and M + 2 exact masses, rather than the RIAs, would be more effective for eliminating incorrect compositions. Because measured RIA values are more subject to interferences than measured exact mass values [18], instances might arise where considering the exact masses might be beneficial for the error limits used above. But for good-quality mass spectra, ignoring the exact masses of the M + 1 and M + 2 mass peaks is a simplification entailing minimal sacrifice.

3.10   ION CORRELATION AND MASS SPECTRAL DECONVOLUTION

Our current research utilizes a Direct Analysis in Real Time (DART) JEOL AccuTOF® oa-TOFMS. The DART is an open-air, surface-sampling ion source that desorbs and ionizes analytes with a heated stream of metastable helium atoms. Elimination of sample extraction, extract clean up, and chromatography decreases the time per sample analysis more than 100-fold. An autosampler [40] and field sample carrier [41] were designed to permit analysis of 1000 cotton swab wipe samples in a single shift by one analyst. This Autosampler/DART/oa-TOFMS instrument should prove useful for rapidly characterizing and monitoring remediation of Superfund or dispersive sites.

A disadvantage of the DART ion source is that composite mass spectra are often obtained. Figures 3.6a, b, and c are mass spectra acquired for a mixture of three compounds containing different sets of heteroatoms. Three cotton swabs were dipped into a methanol solution of the compounds and transported through the 300 °C helium stream in front of the acceptance cone (Orifice 1) into the mass spectrometer. The three mass spectra in Figure 3.6 were acquired at three Orifice 1 (CID) voltages: 15, 40, and 70 V. Swabs dipped into calibrant solutions customized for each Orifice 1 voltage were exposed to the stream immediately after each analyte swab to provide external mass calibration. A data acquisition method was written to record the data at the three CID voltages during a single data acquisition [42].

Generally, the low, moderate, and high CID voltages used to acquire the mass spectra in Figure 3.6 provide mass spectra containing: 1) precursor ions, protonated dimer ions for some analytes, protonated A:B adduct ions for some analyte mixtures, and few, if any, product ions; 2) precursor ions and easily formed product ions; and 3) product ions and lower abundances of precursor ions, respectively. The three possible precursor ions in Figure 3.6a at m/z 182, 265, and 319 are also evident in Figure 3.6b, which confirms that they are precursor ions, because dimeric ions and their products are not observed at a moderate Orifice 1 voltage. No dimeric ions were observed in the low CID voltage spectrum for the three analytes. The m/z 335 ion is an $[M + H + O]^+$ ion related to the m/z 319 ion.

In Figure 3.6c, product ions from all three precursor ions are evident. The ICP is also an ion non-correlation program. Ions that could only be produced from one

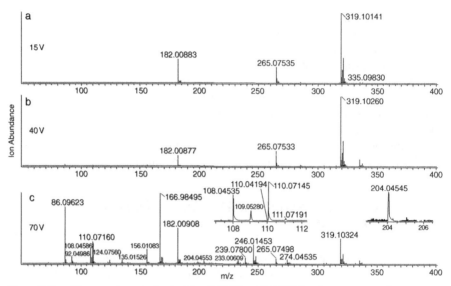

Figure 3.6. Mass spectra acquired with a DART/oa-TOFMS for three compounds on cotton swabs at Orifice 1 (CID) voltages of (a) 15 V, (b) 40 V, and (c) 70 V. No Cl isotopic pattern was observed in the magnified portions of the spectrum for the m/z 108 or 204 ions in (c).

of the three precursor ions will not be correlated with the other two. Table 3.3 lists the ions automatically gleaned by an Ion Extraction Program [42] from text files of m/z ratios and mass peak areas provided by the data system for the three mass spectra in Figure 3.6. The three lists contain monoisotopic precursor and product ions; precursor ions; and oxide, adduct, and dimer ions, if present.

Because RIAs are more susceptible to interferences, the RIAs in Table 3.3 were only calculated and listed when one of two monoisotopic mass peak area thresholds were exceeded. The thresholds were determined empirically from mass spectra obtained for 15 compounds. For the higher threshold, 900 was added to the %RIA to serve as a software flag. During ion correlation, 900 was subtracted from the %RIA and the software applied a 15% RIA error limit; above the lower threshold, the %RIA was unaltered and a 20% RIA was used. Criteria testing with RIAs from only the isotopic mass peaks of the precursor ion and abundant product ions is effective, because once the composition of the precursor ion is limited by RIA criteria, all product ion:neutral loss pairs fragmented from the precursor ion are likewise limited to portions of the precursor ion.

In ref. 19, the ICP first considered the data for the highest mass ion, m/z 319, and then the data for each lower-mass ion sequentially. If no correlation was found, the second ion was also assumed to be a precursor ion to be correlated with other ions later. When a correlation was found, both the m/z 319 ion data and that for the correlated ion were retained and the next lower-mass ion was considered. The previous comparisons were made before the lower-mass ion was considered to ensure

TABLE 3.3.   Exact Masses and RIAs for Ions Gleaned from Text Files of the Mass Spectra in Figure 3.6

Precursor and Product Ions

Exact Mass	%1 RIA	%2 RIA
319.10141	923.42	940.41
265.07535	13.28	6.82
182.00884	910.63	910.00
166.98495	11.41	10.19
86.09623	6.40	0.00
246.01453	13.40	36.06
110.07160	0.00	0.00
156.01084	0.00	0.00
108.04586	0.00	0.00
109.05344	0.00	0.00
135.01526	0.00	0.00
239.07801	0.00	0.00
283.96495	0.00	0.00
297.21216	0.00	0.00
92.04986	0.00	0.00
274.04535	0.00	0.00
124.07560	0.00	0.00
204.04553	0.00	0.00
111.04825	0.00	0.00
233.00609	0.00	0.00
218.07256	0.00	0.00
93.05705	0.00	0.00

Precursor Ions

Exact Mass
319.10141
265.07535
182.00884

Oxides, Adducts, and Dimers

Exact Mass	%1 RIA	%2 RIA	Label
335.09830	0.00	0.00	O

all previously applied constraints on the possible precursor ion compositions were retained. This sequence was repeated until all lower-mass ions were considered or skipped, because they were precursor ions. Table 3.4 illustrates this process for the *m/z* 319 ion.

The current ICP automatically uses multiple sets of mass and RIA error limits to ensure low abundance product ions are not discarded when the error in their measurement slightly exceeds the mass error limit passed by the precursor ion and

TABLE 3.4.   Testing Sequence for Correlating Ions Using the ICP

Nominal Mass	Compositions Found
319	4
319 & 274	2:2:1*
319, 274 & 246	1:1,1:1,1
319, 274, 246 & 239	1:1,1,1:1,1,1
319, 274, 246, 239 & 233	1:1,1,1,1:1,1,1,1
319, 274, 246, 239, 233 & 204	1:1,1,1,1,1:1,1,1,1,1
319, 274, 246, 239, 233, 204 &~~199~~	No compositions
319, 274, 246, 239, 233, 204 &~~167~~	No compositions
319, 274, 246, 239, 233, 204 &~~156~~	No compositions
319, 274, 246, 239, 233, 204 &~~135~~	No compositions
319, 274, 246, 239, 233, 204 &~~124~~	No compositions
319, 274, 246, 239, 233, 204 &~~110~~	No compositions
319, 274, 246, 239, 233, 204 & 109	1:1,1,1,1,1,1:1,1,1,1,1,1
319, 274, 246, 239, 233, 204, 109 &108	1:1,1,1,1,1,1,1:1,1,1,1,1,1,1

*Colons separate the number of compositions found for the precursor ion from those found for the product ions, and those found for the product ions from those found for the neutral losses.

more abundant product ions. The number of possible compositions listed is based on a ±2 mDa mass error range when measured exact masses are accurate to within 2 mDa. More possible compositions may be listed when a ±4 mDa mass error range and ±40% RIA error range are required for a product ion to be recognized as such. More detail on how error limits are automatically adjusted during ion correlation is available in ref. 42. In Table 3.4, no more ions were tested for correlation after seven product ions were found due to memory limitations of QuickBASIC®. The sequence in Table 3.4 was then repeated for the other two precursor ions starting with m/z 265 or m/z 182.

Considering up to seven product ions was satisfactory for individual analytes, but analyte mixtures can provide many more than seven ions. In these cases, this approach sometimes ignored prominent low-mass ions. A software simplification improved the situation. Starting with the most abundant product ion in the product ion mass spectrum and ending with the least abundant ion, each product ion was tested for correlation with each precursor ion. As illustrated in Table 3.5 for a single analyte, metoprolol, the number of possible precursor ion compositions could decrease as more individual product ions were tested for correlation.

Four possible compositions were found for the m/z 268 precursor ion when the elements C, H, N, O, F, P, and S were considered. The most abundant product ion, m/z 116, was correlated with the m/z 268 ion and provided the two possible product ion compositions and four corresponding neutral loss compositions listed in Table 3.5. One possible precursor ion was eliminated. The next most abundant product ion, m/z 191, also correlated with the m/z 268 ion, providing two possible product ion compositions and two corresponding neutral losses. Only two possible precursor ions remained viable. Correlation between the m/z 268 ion and the seventh most abundant product ion (m/z 226) yielded only one precursor ion, one product

TABLE 3.5   Reduction of the Number of Possible Precursor Ion Compositions for Metoprolol When Product Ions Are Considered[*]

Precursor Ion	
268.18933	$C_{11}H_{22}N_7O^+$, $C_{13}H_{28}F_2NS^+$, $C_{13}H_{25}F_3NO^+$, $\mathbf{C_{15}H_{26}NO_3^+}$
Correlations	
268.18933	$C_{11}H_{22}N_7O^+$, $C_{13}H_{25}F_3NO^+$, $\mathbf{C_{15}H_{26}NO_3^+}$
116.10705	$\left(C_4H_{12}N_4^+ + C_7H_{10}N_3O\right)$ or $(C_6H_{14}NO^+ + C_5H_8N_6$, $C_7H_{11}F_3$, or $C_9H_{12}O_2)$
268.18933	$C_{11}H_{22}N_7O^+$, $\mathbf{C_{15}H_{26}NO_3^+}$
191.10638	$(C_{10}H_{13}N_3O^+ + CH_9N_4)$ or $\left(C_{12}H_{15}O_2^+ + C_3H_{11}NO\right)$
268.18933	$\mathbf{C_{15}H_{26}NO_3^+}$
226.14404	$C_{12}H_{20}NO_3 + C_3H_6$
Re-correlations	
268.18933	$\mathbf{C_{15}H_{26}NO_3^+}$
116.10705	$C_6H_{14}NO^+ + C_9H_{12}O_2$
268.18933	$\mathbf{C_{15}H_{26}NO_3^+}$
191.10638	$C_{12}H_{15}O_2^+ + C_{13}H_{11}NO$

* The correct precursor ion composition is in bold type.

ion, and one corresponding neutral loss. This cycle was repeated for all product ions starting with upper elemental limits provided by the single precursor ion composition, and only one possible composition was found for each product ion and neutral loss. After this re-correlation, the constraints on the precursor ion provided by all correlated ions provided constraints for all product ions. When only one precursor ion composition is found for the first correlation, the second cycle is unnecessary and is not performed.

For the three-compound mixture that provided the mass spectra in Figure 3.6, Table 3.6 lists the correlations that were found among the three precursor ions and the product ions produced from them. Twelve product ions were correlated with both the *m/z* 319 and 265 precursor ions. Several low-mass ion compositions were subunits of multiple precursor ions and thus were correlated with more than one precursor ion. Hence, deconvolution of composite mass spectra by applying exact mass and RIA criteria is not as complete as is deconvolution based on small differences in elution times or chromatographic peak shapes for ions when chromatography is used. In Table 3.6, the *m/z* 92, 93, and 135 ions are listed under all three precursor ions, because their compositions are subunits of each one, and which precursor ion(s) produced the ions is indiscernible. The *m/z* 86, 108, 109, 204, and 218 ions correlated with both the *m/z* 319 and 265 precursor ions, but the *m/z* 108 and 204 ions can be ascribed to the *m/z* 265 precursor ion, since the isotopic pattern for a single Cl atom is not evident for these ions in Figure 3.6c. The *m/z* 218 ion is not evident in the spectrum and was an artifact that appeared when the raw data was centroided.

TABLE 3.6.   Ion Correlation Test Results from Three Analytes

Ion Exact Mass	Composition	Mass Error (mDa)	Ion Exact Mass	Composition	Mass Error (mDa)	Ion Exact Mass	Composition	Mass Error (mDa)
319.10141	**C$_{17}$H$_{20}$N$_2$ClS$^+$**	**(–1.6)***	**265.07535**	**C$_{11}$H$_{13}$N$_4$O$_2$S$^+$**	**(–0.0)**	**182.00883**	**C$_8$H$_8$NS$_2^+$**	**(–0.4)**
86.09623	C$_5$H$_{12}$N$^+$	(–0.2)	86.09623	C$_5$H$_{12}$N$^+$	(–0.2)	**166.98495**	**C$_7$H$_5$NS$_2^+$**	**(–0.8)**
	C$_{12}$H$_8$NClS#	(–0.5)		C$_6$HN$_3$O$_2$S	(+0.2)		CH$_3$	(+0.4)
246.01453	**C$_{13}$H$_9$NClS$^+$**	**(+0.7)**	**110.07160**	**C$_5$H$_8$N$_3^+$**	**(+0.3)**	135.01526	C$_7$H$_5$NS$^+$	(+1.5)
	C$_4$H$_{11}$N	(–2.3)		C$_6$H$_5$NO$_2$S	(–0.3)		CH$_3$S	(–2.0)
108.04586	C$_3$H$_9$N$_2$Cl$^+$	(+1.0)		C$_7$H$_{10}$O$^+$	(–1.0)	92.04986	C$_6$H$_6$N$^+$	(+0.4)
	C$_{14}$H$_{11}$S	(–2.6)		C$_4$H$_3$N$_4$OS	(+1.0)		C$_2$H$_2$S$_2$	(–0.8)
109.05344	C$_3$H$_{10}$N$_2$Cl$^+$	(+0.7)	**156.01083**	**C$_4$H$_4$N$_4$OS$^+$**	**(+0.8)**	93.05705	C$_6$H$_7$N$^+$	(–0.3)
	C$_{14}$H$_{10}$S	(–2.4)		C$_7$H$_9$O	(–0.8)		C$_2$HS$_2$	(–0.2)
135.01526	C$_7$H$_5$NS$^+$	(+1.5)		C$_6$H$_6$NO$_2$S$^+$	(–0.5)			
	C$_{10}$H$_{15}$NCl	(–3.2)		C$_5$H$_7$N$_3$	(+0.5)			
239.07800	**C$_{15}$H$_{13}$NS$^+$**	**(+1.7)**	108.04586	CH$_8$N$_4$S$^+$	(–0.6)			
	C$_2$H$_7$NCl	(–3.3)		C$_{10}$H$_6$O$_2$	(+0.5)			
92.04986	C$_6$H$_6$N$^+$	(+0.4)		C$_3$H$_{10}$NOS$^+$	(–1.9)			
	C$_{11}$H$_{14}$NSCl	(–2.0)		C$_8$H$_3$N$_3$O	(+1.9)			
274.04535	**C$_{15}$H$_{13}$NSCl$^+$**	**(+0.2)**		C$_6$H$_6$NO$^+$	(+1.5)			
	C$_2$H$_7$N	(–1.8)		C$_5$H$_7$N$_3$OS	(–1.5)			
204.04553	C$_{11}$H$_9$N$_2$Cl$^+$	(+0.7)	109.05344	CH$_9$N$_4$S$^+$	(–0.8)			
	C$_6$H$_{11}$S	(–2.3)		C$_{10}$H$_4$O$_2$	(+0.8)			
233.00609	**C$_{12}$H$_8$NSCl$^+$**	**(+0.0)**		C$_6$H$_7$NO$^+$	(+1.2)			
	C$_5$H$_{12}$N	(–0.7)		C$_5$H$_6$N$_3$OS	(–1.2)			
218.07256	C$_{13}$H$_{13}$NCl$^+$	(–0.5)	135.01526	C$_7$H$_5$NS$^+$	(+1.5)			
	C$_4$H$_7$NS	(–1.1)		C$_4$H$_8$N$_3$O$_2$	(–1.6)			
93.05705	C$_6$H$_7$N$^+$	(–0.3)	92.04986	C$_6$H$_6$N$^+$	(+0.4)			
	C$_{11}$H$_{13}$NSCl	(–1.4)		C$_5$H$_7$N$_3$O$_2$S	(–0.4)			
			124.07560	**C$_5$H$_8$N$_4^+$**	**(+1.3)**			
				C$_6$H$_5$O$_2$S	(–1.3)			
				C$_7$H$_{10}$NO$^+$	(–0.1)			
				C$_4$H$_3$N$_3$OS	(+0.1)			
			204.04553	C$_9$H$_8$N$_4$S$^+$	(–0.9)			
				C$_2$H$_5$O$_2$	(+0.9)			
			111.04825	**C$_3$H$_{11}$O$_2$S$^+$**	**(+0.8)**			
				C$_8$H$_2$N$_4$	(–0.8)			
			218.07256	C$_{11}$H$_{12}$N$_3$S$^+$	(–2.1)			
				HNO$_2$	(+2.1)			
			93.05705	C$_6$H$_7$N$^+$	(–0.3)			
				C$_5$H$_6$N$_3$O$_2$S	(+0.2)			

*Ions in bold type correlated with only one precursor ion.
#Indented compositions are neutral losses corresponding to the preceding product ion.

To tentatively identify the three compounds, the measured exact masses of the protonated molecules and the RIAs could be compared to calculated values within an exact mass and RIA library. Mass spectra acquired for standards are not required to compile the library; exact masses and RIAs are simply calculated from the compositions of the protonated molecules. For example, several compounds with a molecular weight of 264 g/mole were found in ref. 43. Those compositions with exact masses for the protonated molecules closest to the exact mass of the m/z 265 precursor ion in Figure 3.6 are listed in Table 3.7. The structures of the compounds in Table 3.7 contained at least one primary, secondary, or tertiary amine and would be expected to provide a protonated molecule with ESI, APCI, or DART ionization.

TABLE 3.7.   An Exact Mass and RIA Library for m/z 265[*]

Compound	Composition	Exact Mass	%1 RIA	%2 RIA	ΔM(mDa)[‡]
Sulfadizole	$C_7H_{13}N_4O_3S_2^+$	265.04236	10.91	10.20	
Furazidine	$C_{10}H_9N_4O_5^+$	265.05675	12.59	1.76	14.4
Sulfamerazine	$C_{11}H_{13}N_4O_2S^+$	265.07537	14.40	5.90	18.6
Sulfaperin	$C_{11}H_{13}N_4O_2S^+$	265.07537	14.40	5.90	0.0
Temodox	$C_{12}H_{13}N_2O_5^+$	265.08190	14.06	1.94	6.5
Sulfirame	$C_{10}H_{21}N_2S_3^+$	265.08614	14.20	14.50	4.2
Analyte	$C_{11}H_{13}N_4O_2S^+$	265.07498	13.10	5.28	

[*] Exact masses and RIAs calculated using the atomic masses and isotopic abundances from ref. 21.
[‡] The mass difference between the exact masses of the previous composition and the one in this row.

The measured exact masses for precursor ions from 2-(methylthio)-benzo-thiazole, sulfamerazine, and chlorpromazine, 182.00883, 265.07535, and 319.10141, differed by −0.4, −0.0, and −1.6 mDa, respectively, from their calculated exact masses. In each case, the correct composition would be further investigated in data bases of compounds. For example, in the SciFinder® [20] data base, 14,474; 2,022; and 403 references were found for chlorpromazine, sulfamerazine, and 2-(methyl-thio)-benzothiazole, respectively. For an isomer of 2-(methylthio)-benzothiazole, 3-methyl-2(3H)-benzothiazolethione, 159 references were found and for an isomer of sulfamerazine, Sulfaperin (isosulfamerazine), 138 references. To determine the correct isomers producing the m/z 182 and 265 ions and to confirm the tentative identification of chlorpromazine, all five standards would be purchased and HPLC/ TOFMS would be used to compare retention times and mass spectra acquired with different CID voltages.

3.11   CONCLUSIONS

The measurement of exact masses and RIAs to determine compositions of ions observed in mass spectra has become much simpler over the past decade. Initially, our group plotted mass peak profiles from SIR data acquired with double focusing mass spectrometers. Full scan and three SIR data acquisitions were required, with automatically prepared, analyte-specific SIR menus, which required prior input from the operator. Today, with ESI, APCI, or DART ionization, full scan data acquisitions acquired at different CID voltages using a single-MS-stage oa-TOFMS provide exact masses and RIAs of sufficient accuracy to determine elemental compositions of precursor ions, product ions, and neutral losses without the need to target data collection to a pre-specified list of analyte ions. Concurrently, data interpretation has been streamlined and made conceptually simpler. Initially, exact masses and RIAs for partial mass peak profiles were calculated based on summations of Gaussian distributions calculated for each +1 and +2 ion that contributed to the M + 1 and M + 2 profiles, which could be broadened or partially resolved at 10,000 (10% valley) resolving power. Now, only the measured and calculated RIAs are compared and the individual contributions by +1 and +2 ions to the RIAs are totaled based on full profiles, since a resolving power of 6000 (FWHM) does not cause significant profile broadening. To compensate for the wider mass and RIA error limits of an oa-TOFMS relative to a double focusing instrument, an Ion Correlation Program was developed to restrict the number of possible compositions of precursor ions by correlating the compositions of the product ion:neutral loss pairs with the precursor ions. In addition, the simplicity of an oa-TOFMS provides a lower purchase cost, greater reliability, and a smaller footprint. Less operator expertise is required to acquire data and the time needed to determine ion compositions has been greatly reduced. Final drafts of the authors' work referenced herein are available at http:// www.epa.gov/nerlesd1/chemistry/ice/default.htm.

Notice: The United States Environmental Protection Agency through its Office of Research and Development funded and managed the research described here. It has been subjected to Agency review and approved for publication.

REFERENCES

1. Grange, A.H.; Sovocool, G.W.; Donnelly, J.R.; Genicola, F.A.; Gurka, D.F., Identification of pollutants in a municipal well using high resolution mass spectrometry, *Rapid Commun. Mass Spectrom.*, **1998**, 12, 1161–1169.
2. Grange, A.H.; Brumley, W.C.; Sovocool, G.W., Powerful new tools for analyzing environmental contaminants: mass peak profiling from selected-ion-recording data and a profile generation model, *Am. Environ. Lab.*, **1998**, 10, 1–7.
3. Grange, A.H.; Sovocool, G.W., Mass peak profiling from selected ion recording data (MPPSIRD) as a tool for regulatory analyses, *J. AOAC Int'l.*, **1999**, 82, 1443–1457.
4. Grange, A.H.; Sovocool, G.W., Determination of elemental compositions by high resolution mass spectrometry without mass calibrants, *Rapid Commun. Mass Spectrom.*, **1999**, 13, 673–686.
5. Grange, A.H.; Sovocool, G.W., Identifying Endocrine Disruptors by High Resolution Mass Spectrometry in Analysis of Environmental Endocrine Disruptors. Keith HK, Jones TL, Needham LI, editors. Symposium Series 747, Amer. Chem. Soc. Washington, DC; **2000**. p 133–145.
6. Grange, A.H.; Osemwengie, L.; Brilis, G.; Sovocool, G.W., Ion composition elucidation (ICE): an investigative tool for characterization and identification of compounds of regulatory importance, *Internat. J. Environ. Forensics*, **2001**, 2, 61–74.
7. Grange, A.H.; Genicola, F.A.; Sovocool, G.W., Utility of three types of mass spectrometers for determining elemental compositions of ions formed from chromatographically separated compounds, *Rapid Commun. Mass Spectrom.*, **2002**, 16, 2356–2369.
8. Grange, A.H.; Sovocool, G.W., Identifying compounds despite chromatographic limitations: organophosphates in treated sewage, *LC-GC*, **2003**, 21, 1072–1076.
9. Grange, A.H.; Thomas, P.M.; Solomon, M.; Sovocool, G.W., Identification of compounds in South African stream samples using ion composition elucidation (ICE). 51st ASMS Conf. on Mass Spectrom. & Allied Topics; **2003** June 2–6; Montreal, Ontario. Available at http://www.epa.gov/ nerlesd1/chemistry/ice/sa.htm
10. Grange, A.H.; Sovocool, G.W., Identification of contaminants above a pollutant plume by high resolution mass spectrometry, *Environ. Forensics*, **2007**, 8, 391–404.
11. Snyder, S.A.; Kelly, L.; Grange, A.H.; Sovocool, G.W.; Snyder, E.M.; Giesy, J.P., Pharmaceuticals and Personal Care Products in the Waters of Lake Mead, Nevada, in Pharmaceuticals and Personal Care Products in the Environment: Scientific and Regulatory Issues. Daughton CG, Jones-Lepp T, editors. Symposium Series 791, Amer. Chem. Soc. Washington, DC; **2001**. p 116–141.
12. Brumley, W.C.; Grange, A.H.; Kellliher, V.; Patterson, D.B.; Montcalm, A., Environmental screening of acidic compounds based on capillary zone electrophoresis/laser-induce fluorescence detection with

identification by gas chromatography/mass spectrometry and gas chromatography/high resolution mass spectrometry, *J. AOAC Int'l.*, **2000**, 83, 1059–1067.

13. Bobeldijk, I.; Stoks, P.G.M.; Vissers, J.P.C.; Emke, E.; van Leerdam, J.A.; Muilwijk, B.; Berbee, R.; Noij, Th.H.M.J., Surface and wastewater quality monitoring: combination of liquid chromatography with (geno)toxicity detection, diode array detection and tandem mass spectrometry for identification of pollutants, *J. Chromatogr. A*, **2002**, 970, 167–181.

14. Thurman, E.M.; Ferrer, I.; Fernández-Alba, A.R., Matching unknown empirical formulas to chemical structure using LC/MS TOF accurate mass and database searching: example of unknown pesticides on tomato skins, *J. Chromatogr. A.*, **2005**, 1067, 127–134.

15. Ibáñez, M.; Sancho, J.V.; Pozo, O.J.; Niessen, W.; Hernández, F., Use of quadrupole time-of-flight mass spectrometry in the elucidation of unknown compounds present in environmental water, *Rapid Commun. Mass Spectrom.*, **2005**, 19, 169–178.

16. NIST/EPA/NIH Mass Spectral Library, Version 2.0, Standard Reference Data Program, National Institute of Standards and Technology, U.S. Dept. of Commerce, NIST Mass Spectrometry Data Center, Gaithersburg, MD 20899, **2005**.

17. Wiley Registry of Mass Spectra, 6th ed. Palisade Corp., Newfield, NY.

18. Grange, A.H.; Winnik, W.; Ferguson, P.L.; Sovocool, G.W., Using a triple quadrupole mass spectrometer in accurate mass mode and an ion correlation program to identify compounds, *Rapid Commun. Mass Spectrom.*, **2005**, 19, 2699–2715.

19. Grange, A.H.; Zumwalt, M.C.; Sovocool, G.W., Determination of ion and neutral loss compositions and deconvolution of product ion mass spectra using an orthogonal acceleration, time-of-flight mass spectrometer and an ion correlation program, *Rapid Commun. Mass Spectrom.*, **2006**, 20, 89–102.

20. Chemical Abstracts Service, American Chemical Society, Columbus, OH. http://www.cas.org/SCIFINDER/scicover2.html

21. NIST Physics Laboratory. Atomic weights and isotopic compositions for all elements. Available at http://physics.nist.gov/cgi-bin/Compositions/stand_alone.pl?ele=

22. EPA, Method 1613, Revision B, Tetra- through Octa-Chlorinated Dioxins and Furans by Isotope Dilution HRGC/HRMS at http://www.epa.gov/waterscience/methods/1613.pdf

23. EPA, Method 8290, Polychlorodibenzodioxins (PCDDs) and Polychlorodibenzofurans (PCDFs) by High-Resolution Gas Chromatography/High-Resolution Mass Spectrometry (HRGC/HRMS) at http://www.epa.gov/epaoswer/hazwaste/test/pdfs/8290.pdf

24. Grange, A.H.; Donnelly, J.R.; Brumley, W.C.; Billets, S.; Sovocool, G.W., Mass measurements by an accurate and sensitive selected-ion-recording technique, *Anal. Chem.*, **1994**, 66, 4416–4421.

25. Grange, A.H.; Donnelly, J.R.; Brumley, W.C.; Sovocool, G.W., Determination of elemental compositions from mass peak profiles of the molecular (M), M + 1, and M + 2 Ions, *Anal. Chem.*, **1996**, 68, 553–560.

26. Grange, A.H.; Brumley, W.C., A mass peak profile generation model to facilitate determination of elemental compositions of ions based on exact masses and isotopic abundances, *J. Amer. Soc. Mass Spectrom.*, **1997**, 8, 170–182.

27. EPA, **1990** HPV Challenge. U.S. EPA Office of Pollution Prevention and Toxics (Revised May 28, 2004). At http://www.epa.gov/chemrtk/.

28. EPA, Endocrine Disruptor Screening Program. U.S. EPA office Of Science Coordination and Policy. EDSTAC Final Report, Executive Summary, **1998**, P. ES-3. Available at http://www.epa.gov/scipoly/oscpendo/pubs/edstac/exesum14.pdf.

29. Grange, A.H.; Brumley, W.C., Plotting mass peak profiles from selected ion recording data, *Rapid Commun. Mass Spectrom.*, **1992**, 6, 68–70.

30. Grange, A.H.; Brumley, W.C., Mass spectral determination, *Environ. Testing & Analysis*, **1996**, March/April:22–26.

31. Grange, A.H.; Brumley, W.C., Determining elemental compositions from exact masses and relative abundances of ions, *Trends Anal. Chem.*, **1996**, 15, 12–17.

32. Grange, A.H.; Brumley, W.C., Identification of ions produced from components in a complex mixture by determination of exact masses and relative abundances using mass peak profiling, *LC-GC*, **1996**, 14, 478–486.

33. Ferrer, I.; Thurman, E.M., Importance of the electron mass in the calculations of exact mass by time-of-flight mass spectrometry, *Rapid Commun. Mass Spectrom.*, **2007**, 21, 2538–2539.

34. Vetter, W.; Alder, L.; Palavinskas, R., Mass spectrometric characterization of Q1, a C9H3Cl7N2 contaminant in environmental samples, *Rapid Commun. Mass Spectrom.*, **1999**, 13, 2118–2124.

35. Wu, J.; Vetter, W.; Gribble, G.W.; Schneekloth, J.S. Jr.; Blank, D.H.; Görls, H., Structure and synthesis of the natural heptachloro-1'-methyl-1,2'-bipyrrole (Q1), *Angew. Chem. Int. Ed* 41, **2002**, 10, 1740–1743.

36. Jörundsdóttir, H.; Norström, K.; Olsson, M.; Pham-Tuan, H.; Hühnerfuss, H.; Bignert, A.; Bergman, A., Temporal trend of bis(4-chlorophenyl) sulfone, methylsulfonyl-DDE and -PCBs in Baltic Guillemot (Uria aalge) egg 1971–2001—a comparison to 4,4'-DDE and PCB trends, *Environ. Pollut.*, **2006**, 141, 226–237.

37. Suzuki, S.; Ishii, T.; Yasuhara, A.; Sakai, S., Method for the elucidation of the elemental composition of low molecular mass chemicals using exact masses of product ions and neutral losses: application to environmental chemicals measured by liquid chromatography with hybrid quadrupole/time-of-flight mass spectrometry, *Rapid Commun. Mass Spectrom.*, **2005**, 19, 3500–3516.

38. Ojanperä, S.; Pelander, A.; Pelzing, M.; Krebs, I.; Vuori, E.; Ojanperä, I., Isotopic pattern and accurate mass determination in urine drug screening by liquid chromatography/time-of-flight mass spectrometry, *Rapid Commun. Mass Spectrom.*, **2006**, 20, 1161–1167.

39. Kaufmann, A., Determination of the elemental composition of trace analytes in complex matrices using exact masses of product ions and corresponding neutral losses, *Rapid Commun. Mass Spectrom.*, **2007**, 21, 2003–2013.

40. Grange, A.H., An inexpensive autosampler to maximize throughput for an ion source that samples surfaces in open air, *Environ. Forensics*, **2008**, 9, 127–136.

41. Grange, A.H., An integrated wipe sample transport/autosampler to maximize throughput for a DART®/oa-TOFMS, *Environ. Forensics*, **2008**, 9, 137–143.

42. Grange, A.H.; Sovocool, G.W., Automated determination of precursor ion, product ion, and neutral loss compositions and deconvolution of composite mass spectra using ion correlation based on exact masses and relative isotopic abundances, *Rapid Commun. Mass Spectrom.*, **2008**, 22, 2375–2390.

43. Pfleger, K.; Maurer, H.H.; Weber, A., Mass Spectra and GC Data of Drugs, Poisons, Pesticides, Pollutants and Their Metabolites. 2nd edition. New York: VCH; **1992**.

CHAPTER 4

EXISTENCE AND USE OF ACCURATE MASS DATABASES AND ELEMENTAL COMPOSITION TOOLS FOR TARGET AND NON-TARGET ANALYSES

Imma Ferrer and E. Michael Thurman

Center for Environmental Mass Spectrometry, Department of Civil, Environmental and Architectural Engineering, University of Colorado, Boulder, Colorado

TRADITIONALLY, the database screening of environmental contaminants has been accomplished by GC/MS methods using conventional library searching routines. However, many of the new polar and thermally labile compounds are more readily and easily analyzed by LC/MS methods and no searchable libraries currently exist (with the exception of some user libraries, which are limited). Therefore, there is a need for LC/MS libraries that can detect organic compounds such as pesticides and pharmaceuticals and their degradation products. This chapter illustrates two different approaches where accurate mass databases have been developed and applied to the analysis of pesticides in environmental samples, such as food and water. The first approach is a tailor-made database that uses Microsoft Office Access for 350 pesticides that are amenable to detection by LC/TOF-MS. Accurate masses and empirical formula of compounds detected by time-of-flight are matched against the database for a positive confirmation. The second approach involves a more sophisticated technique that uses an automated molecular feature database (MFD) which compiles accurate mass ions for a set of known compounds. Furthermore, the combined use of accurate mass and chromatographic retention time eliminates false positives in the automated analysis used in this approach.

Liquid Chromatography Time-of-Flight Mass Spectrometry: Principles, Tools, and Applications for Accurate Mass Analysis, Edited by Imma Ferrer and E. Michael Thurman
Copyright © 2009 John Wiley & Sons, Inc.

4.1   INTRODUCTION

Usually, methods that use gas chromatography/mass spectrometry (GC/MS) apply a reverse-search approach that consists of matching the spectra obtained to the large National Institute for Standards and Testing (NIST) pesticide libraries in minutes [1]. This has made screening quite simple for pesticides amenable to GC/MS. Unfortunately, similar reverse-search methods have not been available for liquid chromatography/mass spectrometry (LC/MS). Single quadrupole and triple quadrupole mass spectrometers do not operate in full scan mode for pesticide screening because of a lack of sensitivity [2, 3]. Although libraries for LC/MS three-dimensional ion trap and triple quad have been developed, they have not been popular due to difficulties in reproducibility of fragmentation and the need for authentic standard analysis for each instrument [4–6]. Recently liquid chromatography/quadruple/ion trap (LC/Q-trap with a linear ion trap) and automated library searching has been used for drugs in blood and urine and screening of 301 drugs has been reported [7]. There is considerable interest in this technique because of the combination of both multiple reaction monitoring (MRM) and full scan spectra being taken by what is called "a data dependent scan" (Applied Biosystems, Foster City, CA). However, with LC/Q-trap (as with triple quadrupole instrumentation) there is a limitation to the number of ions that may be selected because of dwell time and ion windows.

The reason why liquid chromatography/time-of-flight mass spectrometry (LC/TOF-MS) has become popular for identification of non-target compounds is that it uses full scan information, which is both sensitive and accurate [8–10]. The combination of accurate mass and sensitivity is needed for screening of pesticides by their empirical formula. For example, Thurman et al. [11] used an approach of TOF, ion trap, and the Merck Index database to identify pesticides in food and also degradation products, without the initial use of primary standards [12]. Bobeldijk et al. also used the Merck Index, the NIST library, and their own database to screen water pollutants [13]. The methods in these examples rely on manually searching the databases, compound by compound. Recently, however, several papers have appeared by Laks et al. [14, 15] that use mass accuracy of 30 ppm and database analysis to identify ~600 drugs in blood and urine without the use of primary standards, using only the protonated molecule. The authors used an automated data explorer program and LC/TOF-MS.

In spite of the progress that has been made, the ability to do true library analysis is still a problem to be solved for liquid chromatography/mass spectrometry and for rapid analysis of environmental samples. The problems to be overcome include reproducible spectra and ion ratios, routine programs for rapid screening of samples rather than manual checking of data, and some estimate of the probability of the correct identification. Variation in fragmentation intensity is not critical with the use of accurate mass since the accurate mass of the fragment ion gives its molecular formula. The database approach is a screening tool and it is powerful and fast because only the molecular formula is needed.

In this chapter, the contribution to the solution of these problems is presented using two different approaches. The first one uses a user-created simple database that uses Microsoft Office Access and accurate masses for a large number of pesticides. The second approach uses an automated molecular feature database that allows the user to screen and identify 600 pesticides in food and water extracts using positive ionization LC/TOF-MS and the monoisotopic exact mass of the MH$^+$, at least one product ion, and retention time of the compound. The advantages and limitations of both approaches are discussed as well as the reliability (match probability) of a library search using accurate mass.

4.2 EXPERIMENTAL

The separation of the different pesticides was carried out using an HPLC system (consisting of vacuum degasser, autosampler, and a binary pump) (Agilent Series 1100, Agilent Technologies, Palo Alto, CA) equipped with a reversed-phase C_8 analytical column of 150 mm × 4.6 mm and 5 μm particle size (Zorbax Eclipse XDB-C8). Column temperature was maintained at 25 °C. Mobile phases A and B were water with 0.1% formic acid and acetonitrile, respectively. A binary gradient elution was made as follows: isocratic conditions for 5 min at 10% of solvent B, then linear gradient from 10 to 100% of solvent B, from 5 to 30 min. The flow rate used was kept at 0.6 mL min^{-1} and 50 μL of sample extracts were injected in each study. This HPLC system was interfaced to a time-of-flight mass spectrometer Agilent MSD TOF (Agilent Technologies, Palo Alto, CA) equipped with an electrospray interface operating in positive ion mode, using the following operation parameters: capillary voltage: 4000 V; nebulizer pressure: 40 psig; drying gas: 9 L min^{-1}; source temperature: 300 °C; fragmentor voltage: 190 V and 230 V; skimmer voltage: 60 V; octapole DC 1: 37.5 V; octapole RF: 250 V. LC/MS accurate mass spectra were recorded across the range m/z 50–1000. The instrument provided a typical resolving power (FWHM) of 9500 ± 500 (m/z 922.0098). The full-scan data recorded was processed with Applied Biosystems/MDS-SCIEX Analyst QS software (Frankfurt, Germany) with accurate mass application-specific additions from Agilent MSD TOF software. The molecular feature search is a recent addition (2005) of the Agilent TOF software.

Accurate mass measurements of each peak from the total ion chromatograms were obtained using a dual-nebulizer electrospray source with an automated calibrant delivery system, which introduces the flow from the outlet of the chromatograph together with a low flow of a calibrating solution containing the internal reference compounds purine and *hexakis(1H, 1H, 3H-tetrafluoropropoxy)phosphazine*, respectively (masses m/z 121.0509 and 922.0098).

4.3 RESULTS AND DISCUSSION

4.3.1 Microsoft Office Access Database

Concept of the Database   The concept here consists of three parts. First is the initial screening of possible pesticides in real vegetable and fruit extracts using accurate mass and generating an accurate mass via an automatic ion extraction routine with the LC/TOF-MS instrument. Second is searching the Access database manually for a screening identification of a pesticide. Third is identification of the suspected compound by accurate mass of at least one fragment ion and comparison of retention time with an actual standard. The combination of LC/MS-TOF and database screening can be summarized then in three steps:

1. Analyze the fruit or vegetable extract with LC/TOF-MS in full scan looking for important unknown peaks using a mild in-source CID fragmentation.

2. Search the Access pesticide database using the accurate mass found in full scan for initial screening.

3. If samples screen positive for a pesticide, make identification using at least one accurate mass fragment ion combined with retention time.

Development of the Database   The database was manually created by inclusion of the formula and accurate masses for 350 different pesticides. The list consists of the three major classes of pesticides that are commonly used in Europe for treating crops of fruits and vegetables, as well as many of the pesticide degradation products. There are insecticides, herbicides, and fungicides. An Excel spreadsheet was used to calculate the accurate masses of each of the 350 pesticides. The compounds in the database are known to ionize in positive ion electrospray and were collated from a series of publications, as well as from approximately 100 standards analyzed in our laboratory. Figure 4.1 shows the appearance of the Access database. The front page of the database has four basic key functions. The first is called "Find Accurate Mass" and consists of the accurate masses of all 350 pesticides to the fourth decimal place. The program searches these and finds all compounds that have the correct mass to the second decimal place. This allows searching for compounds that have either similar or identical accurate masses. It turns out that, although accurate mass is a very powerful technique, it is possible for pesticides to have the same empirical

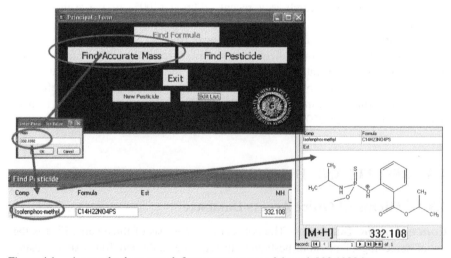

Figure 4.1.   Access database search for accurate mass of the *m/z* 332.1080 ion.

formula and thus the same accurate mass! In fact in this database of 350 compounds, 8 compounds have the same formula. This includes several degradation products of the triazines. These compounds also require retention time and fragment ions for a final confirmation.

The second key of the database is "Find Pesticide." This key will locate any compound in the database and display its formula, structure, and accurate mass to four decimal points. The third key is "New Pesticide" key, which is used to add new compounds to the database. The fourth key is "Edit List," which allows one to change parameters of the compounds within the database. The edit key is disabled when the program is used in the laboratory in order to protect the safety of the program from error entries. The fifth and last key is the "Exit" key, which exits the database program.

Advantages of using Access for this pesticide database is that this program is available to use on any computer that runs Windows. The database is approximately 18 megabytes in size; thus, it is easily transported and available for use by other scientists that are working on pesticide analysis. It can be readily sent by email or transferred by memory stick, CD, etc. The database is used in manual mode by entering the accurate mass; this is a limitation at this time.

Example of Identification of Isofenphos-Methyl in Pepper Extracts
Figure 4.2 shows the total ion chromatogram (TIC) for the extraction of a store-purchased pepper. There are approximately 20 major peaks in the chromatogram at retention times from 2 to 30 minutes. However, the matrix is not simple and over 1000 peaks are present in sample with a signal to noise of $10:1$. However, in this manual application we are screening only the major peaks. As an example, the peak at 27.2 minutes was screened as a possible pesticide by the manual database search. The accurate mass for this peak was m/z 332.1082. This ion was chosen because it is the protonated molecule of the peak at 27.2 minutes. The database gave a positive screen for isofenphos methyl (Figure 4.1). Recent rapid sanitary information or alerts reported by European countries have pointed out a serious problem related to the presence of illegal or misused pesticides, such as isofenphos methyl, particularly in peppers. Here we differentiate between a screen and an identification of a pesticide.

Screening is a test that shows that a possible pesticide is present based on an analysis that is inadequate for confirmation, but that is rapid and inexpensive (i.e., the database search of the molecular ion with a retention time). The identification is based on further analysis, which includes the accurate mass of the protonated molecule and one fragment ion with the correct retention-time match.

Thus, continuing with the example of a positive screen, the database works by first entering the accurate mass of the protonated molecule. The mass is checked versus the database to two decimal places (note the value of 332.11 shown in Figure 4.1). Next the compound name is entered in the database and the structure, formula, and accurate mass appear for isofenphos-methyl. The accuracy in this case was a difference of 0.2 mDa or ~0.6 ppm. This procedure completes the screen for isofenphos-methyl, but the compound is not yet identified. Identification requires more information, including fragment ion and retention time matching, which will be addressed as follows.

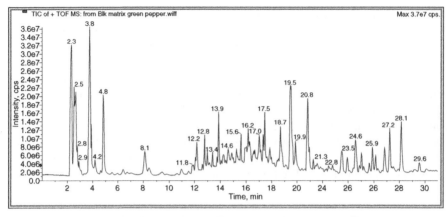

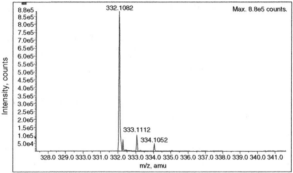

Figure 4.2. Total ion chromatogram of a pepper sample and accurate mass spectrum for peak at 27.2 minutes.

The usual approach to identify the compounds is to acquire the standard and to measure both retention time and at least one fragment ion for identification [16]. The accurate mass fragment ions for isofenphos are m/z 273.0345 which corresponds to $[C_{11}H_{14}O_4PS]^+$ and m/z 230.9875 corresponding to $[C_8H_8O_4PS]^+$. In both cases the fragments were identified by the characteristic isotopic signature of the sulphur atom. According to the identification point system [2, 16], the high resolution accurate mass of the protonated molecule and the fragment ion give a total of 4 IPs, which is used for identification along with the ion ratio. Nonetheless, further information can also be obtained from the characteristic sulphur isotope signature of both the protonated molecule and the fragment ions in this case.

Example of Identification of DEET in Surface Water Samples Another example is shown in Figure 4.3 for the analysis of a surface water sample from Kansas. In this case we focused on the peak at 21.3 minutes. The experimental accurate mass obtained for this peak was 192.1382. Following the same procedure carried out in the previous section, the database match for this accurate mass was DEET, which is an insect repellent pesticide, commonly used in the United States. Thus, identification

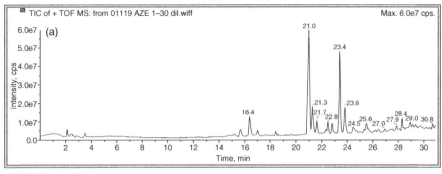

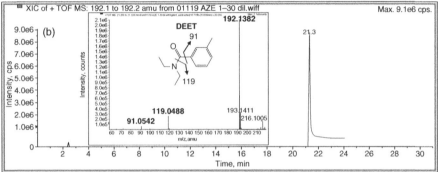

Figure 4.3. (a) Total ion chromatogram corresponding to the LC/TOF-MS analysis of a surface water sample where DEET was detected. (b) Extracted ion chromatogram of DEET (inset: accurate mass spectrum).

of this compound was performed by injection of a standard and a close look at the fragment ions.

Note that this three-step procedure requires first the screening of the pesticide followed by an identification step. The goal of this project, ultimately, was to include this information in the database, both fragment ion and retention time, and to automate the process. And this is what the second approach shown in the next section will describe.

4.3.2 Molecular Feature Database

Concept of the Database   The algorithm of the database search is called a molecular feature extractor, which finds all ions that represent real compounds eliminating all ions that are background noise or noise spikes. The search software then compares the ions found to all specified adducts in the compound database. The user specifies the adducts sought (proton, NH_4^+, Na^+, etc.). The molecular feature approach is more suited to large databases because of the ease of operation and the speed for which the search is done compared to the extracted ion approach, which is much slower. In contrast, a reverse-search database extracts each ion of interest in the database from the sample file and compares the accurate mass with the database, which is a time-consuming process (about 10–100 times longer), but is how one typically analyzes the data by hand.

The creation of the database library involves three steps: selection of the pesticide and its product ion, calculation of the exact masses of the molecules, and the creation of the csv (comma separated values) file. A description of each molecule and its retention time may be included but is not necessary for database operation. The csv file is then the database that is searched. The process involves the selection of a method file that uses the LC/TOF-MS sample data-file, which has already been analyzed on the instrument, and the actual processing of the work list to generate the MFD search and data report. Typically on the most complex food sample (the pepper extract of ~2500 peaks, Figure 4.2), a database of 100 compounds is searched in 2 to 5 minutes, including the printed report.

Development of the Database   Theoretical monoisotopic exact masses of compounds based on their molecular formula were calculated using an Excel spreadsheet and put into csv format for use by the instrument software of the LC/TOF-MS system. The LC/TOF-MS instrument software at the completion of the sample run then searches the csv file automatically and generates a report on the number of compounds that screened positive in the database. Search criteria include ppm mass tolerance (5 ppm), retention time window (0.2 minutes), and minimum signal (1000 counts or a signal to noise of ~10:1) or other variables chosen by the operator including adducts and neutral loss fragments. A minimum of one accurate mass product ion is also included in the database for identification. The csv file may be created by use of a spreadsheet supplied by the instrument software that calculates the exact mass of the molecule of interest. One enters the number of carbon, hydrogen, oxygen, nitrogen, chlorine, or sulfur atoms and the spreadsheet calculates the exact mass of the neutral compound. The specified adduct (H^+, NH_4^+ etc) is then used for the accurate mass comparison of the MFD search with an ± error of 5 ppm (set by the user).

A file containing 600 pesticides was obtained by calculating accurate masses for the protonated molecules and product ions. Retention time matching was also used with an error of ±0.01 minutes via a deconvolution software program (called Molecular Hunter). These compounds consist of the major pesticides that are used in Europe and in the United States for treating crops of fruits and vegetables and those compounds commonly reported in water quality studies [17]. They include a mixture of insecticides (organophosphates, carbamates, and imidazoles), herbicides (triazines and phenylureas), and fungicides (imidazoles and thiazoles).

Example of Identification of Pesticides in Vegetable and Fruit Samples
Table 4.1 shows the results of a library search of 6 fruit and vegetable samples from a local grocery store (apple, pear, tomato, potato, pepper, cucumber) and one commercial brand of olive oil for the 600 pesticides in the database. The MFD search found from 617 to 2681 accurate mass peaks in the sample chromatograms (Table 4.1). The least complicated sample matrix was the tomato with 617 peaks, and the apple and pepper were the most complex samples with ~2500 peaks each. The sensitivity of the MFD search was set at a signal-to-noise of 10:1. The quantity of peaks found approximately doubles with decreasing the signal-to-noise ratio from 20:1 to

TABLE 4.1. Screened Pesticides in Food and Water Samples Using the MF Database

Sample	Peaks Screened	Pesticide Matches <5 ppm	Pesticides Identified LC/TOF MS	Error mDa	Error ppm	Retention Time Error (min.)	False Neg.	False Pos.
Apple	2681	12	Imazalil				1	0
			Imazalil degradate	−1	−3.9	−0.08		
			Iprodione	0.22	0.7	−0.12		
			Fluquinconazole	0.05	0.1			
			Difenoconazole	−0.74	−1.8			
Olive Oil	1678	10	Terbuthylazine	0.06	0.2	0.09	0	0
			(Deisopropylatrazine)					1
Pepper	2402	41	Imazalil	0.3	1	0.07	0	
			Diazinon	0.11	0.3	0.12		
			Buprofezin	1	3.3	0.1		
Tomato	617	8	Buprofezin				3	
			Carbendazim					
			Thiophanate methyl					
Cucumber	1619	17	Thiabendazole	0.2	1	0.01	0	0
			Malathion isomer 1	0.04	0.1	0.05		
			Malathion isomer 2	0.22	0.7	−0.03		
			Malathion oxon	0.25	0.8	0.08		
			Imazalil	−0.1	0.3	0.05		
Pear	1209	14	Imazalil				1	0
			Carbendazim	−0.21	−1.1	0.05		
			Imazalil degradate	0.51	2	0.03		
			Phosmet	0.31	1	0.04		
Potato	1150	11	None				0	0
Surface Water Kansas	1271	25	Atrazine				1	
			Prometon	0.02	0.1	−0.02		
			Hydroxyatrazine	−0.35	−1.8	−0.04		
			Deisopropylatrazine	−0.15	−0.9	−0.05		
			Simazine	−0.44	−2.2	−0.04		
			Diuron	0.1	0.4	−0.04		
			Propazine	−0.18	−0.8	−0.01		
			Dimethenamide	−0.17	−0.6	−0.04		
			Metolachlor	0.07	0.2	−0.04		
			(Propachlor)					1
Surface Water Kansas	1710	25	Atrazine				2	
			Terbuthylazine					
			Deisopropylatrazine	−0.41	−2.4	−0.5		
			Simazine	−0.33	−1.7	−0.05		
			Diuron	−0.52	−2.3	−0.04		
			Propazine	−0.35	−1.5	−0.01		
			Dimethenamide	−0.03	−0.1	−0.04		
			Metolachlor	0.1	0.3	−0.04		
			Bromacil	−0.56	−2.2	0.02		
			Diazinon	0.34	1.1	−0.13		
			Deethylatrazine	−0.35	−1.9	−0.08		
			Absolute Means	**0.30**	**1.18**	**0.07**		
			Absolute Std Dev	**0.25**	**0.97**	**0.09**		

10:1. The value of 10:1 was chosen in order to obtain good X + 1 and X + 2 isotope signatures of the compound with the maximum instrument sensitivity.

The accuracy window of the MFD search was set at 5 ppm in order to be well within the mass accuracy of the LC/TOF-MS system. The number of pesticides found in the 5-ppm mass window of these samples varied from 8 to 41 compounds. The only criterion to be included in this match was that the MH⁺ ion was within 5-ppm of the database value. Thus, as an example, the pepper sample, which contained 2402 peaks, had only 41 of these peaks that met the 5-ppm accuracy window (Table 4.1). Of these 41 peaks, only three formulas were identified based on the correct isotope signature, and the correct retention-time match. They were the compounds imazalil, diazinon, and buprofezin. The identification was checked not only in the printout of the automated database match but also by manual confirmation of the data file.

The confirmation of the empirical formula varied from no detections in the potato sample, one pesticide in olive oil, three pesticides in pepper and tomato, and five pesticides in the cucumber and apple. The most common compound identified by the screen in the fruit and vegetable samples was imazalil, which is a post-harvest fungicide used for transport and storage of fruits and vegetables before their sale. Other compounds included organophosphate insecticides, such as diazinon, phosmet, and malathion and the oxon of malathion, which is a pesticide degradation product. The insect growth regulator buprofezin was found in a tomato and pepper sample.

The tomato sample was most problematic for the screening database, and in some ways, the most useful sample analyzed. It contained three false negatives (i.e., screened negative but contained buprofezin, thiophanate methyl, and carbendazim). The cause of the false negatives was the high concentration (greater than 1 mg/kg) of the pesticides present in the tomato, which skewed the accurate mass outside of the 5-ppm window due to partial saturation of the detector. This problem may be corrected by using the X + 1 ^{13}C isotope of the pesticide for screening, which kept the mass accuracy within the 5-ppm accuracy window. The same problem of false negatives occurred with the pear sample for imazalil. This was corrected also by the addition of the X + 1 isotope of imazalil. Thus, the database has now been corrected for all of the major high use compounds by the addition of the ^{13}C isotope of the MH⁺ ion. This correction is done by spreadsheet and the subtraction of 12.00000 for one carbon atom and the addition of 13.00335 for the accurate mass of the ^{13}C atom. This correction does not take into account other X + 1 isotopes (i.e., ^{2}H or ^{15}N), so it is a close approximation (within 0.5 mDa) meant to find extremely high concentrations of parent pesticide only but does easily stay within a mass accuracy of 5 ppm. Approximately an additional error of 0.1 to 1 ppm is added by using the ^{13}C isotope only and is dependent on the number of carbon, hydrogen, and nitrogen atoms in the compound. However, new recent developments in the software take into account this problem of high concentrations and correct the accurate mass of the analyte.

The accuracy of all confirmed samples had an absolute-value average of 0.3 mDa or ~1.2 ppm and a standard deviation of 0.25 mDa and ~1.0 ppm, respectively (Table 4.1). The absolute-value average for retention time match was 0.07 minutes and standard deviation of 0.09 minutes. Thus, the windows chosen for the database search are chosen with enough margin of error to find 99% of the pesticides based on two standard deviations of the mean for mass accuracy and retention time.

Example of Identification of Pesticides in Surface Water Samples   The MFD search was tested for extracts of two water samples taken during the spring runoff [17] in the state of Kansas. This time of year is important for the runoff of herbicides from corn, wheat, and sorghum that are treated with herbicide prior to planting in April and May [17]. The MFD search found 1271 peaks in the first sample, of those, 25 peaks fell within the 5-ppm accuracy window and 9 formulas were confirmed by accurate mass, isotopic ratio, and retention-time match. This sample contained also one false negative, atrazine, because of the extremely high concentration of this compound (5 mg/kg in the methanol extract), which skewed the mass accuracy. This problem was corrected, again, in the same manner as for the food samples, which is by the addition of the ^{13}C peak for atrazine at m/z 217.1041.

A false positive occurred for the compound propachlor at m/z 212.0836, which is also the same empirical formula and same exact mass as a product ion of metolachlor. The X + 2 isotopic signature is correct of course because propachlor and the metolachlor product ion have the same formula. The retention time was not available for propachlor but was for metolachlor. A simple fragmentation study of metolachlor at higher fragmentor voltages did show that the m/z 212.0836 ion originated from metolachlor as a low intensity product ion.

An example printout of the MFD is shown in Figure 4.4. Diuron, a common herbicide used in crops was detected in the surface water sample as shown in this figure. The report contains formula, compound, exact mass of the neutral molecule, error in mDa and error in ppm, retention time error in minutes, and a description (i.e., diuron, herbicide). The mass spectrum of the compound shows the MH$^+$ and isotope signature of the compound, which is useful for a quick check and partial

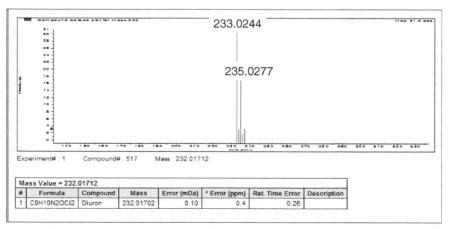

Figure 4.4.   Example of a report from an MFD search; mass of neutral is reported in the printout and MH+ is shown with the X + 2 ion cluster.

confirmation of the formula, especially since most pesticides (about 70% of the 100 compound database) show an interesting X + 2 ion from a halogen or sulfur atom.

The second water sample contained 1710 peaks of which 25 were screened under 5-ppm mass accuracy window. Of these, 11 formulas were confirmed. There were two false negatives caused by high concentrations, atrazine and terbuthylazine. Both of these compounds were detected by the use of the ^{13}C isotope (error within 2 ppm). Terbuthylazine was used in this sample as an internal standard. The solid-phase extraction, which multiplied the concentration in water by 300 fold, created the over saturation for this compound and the loss of mass accuracy. Manual screening of the large peaks was used to detect these false negatives.

These experiments indicate that the MFD search with monoisotopic mass only (and X + 2 confirmation) is useful for data exploration and is a quick way to search for many compounds of interest. The MFD was useful to find compounds such as fluquinconazole and difenoconazole in apples, and hydroxyatrazine, prometon, and diuron in water samples, which were not seen by the first manual inspection of the sample file.

In summary, the MFD software compiles accurate mass ions, excludes noise, and compares them with the monoisotopic exact masses of the database. The screening criteria consisted of ±5 ppm accurate mass window, ±0.2 minute retention time window, and a minimum 1000 counts (signal to noise of ~10 : 1). The MFD search was useful for rapid screening and identification of pesticides in food and water as shown in actual samples. The combined use of accurate mass and chromatographic retention time eliminated false positives in the automated analysis. The major weakness of the MFD is matrix interferences and loss of mass accuracy. Strengths of the MFD include rapid screening of 600 compounds at sensitive levels compared to a manual approach and the ease of use of the library for any accurate mass spectrometer instrumentation capable of routine sub 5-ppm mass accuracy.

4.4   CONCLUSIONS

In recent years there has been an important need for research studies and methods development on the analysis and screening of pesticides in food by new LC/MS methods, such as accurate mass screening using LC/TOFMS and database analysis. This combined with the use of large accurate mass databases of pesticides is an important and useful new screening procedure, especially when accurate mass is less than 3-ppm accuracy. Thus, the use of accurate mass LC/TOFMS combined with database searching is a powerful example of a new wave of technology for screening of pesticides in food and environmental samples.

Databases can be manually created with simple programs such as Microsoft Access, or easily automated like the MFD approach. The strength of the MFD search is the speed of screening hundreds of pesticides in minutes, including both their protonated molecules and product ions. The use of classical fragmentation libraries with comparison to fragmentation patterns is not needed in LC/TOF-MS using accurate mass because the match is based on presence of the ion rather than the intensity of the ion and its fragmentation pattern. The problem of instrument variation and matrix effects on fragmentation is partially eliminated with accurate mass. However, it may be necessary to monitor more than one product ion depending on the type of pesticide being determined. Pesticides without characteristic X + 2 isotopic patterns need at least two product ions.

It is the view of the authors that a large problem in LC/MS libraries is on the verge of being solved with the use of accurate mass databases for pesticide screening in food and water using the protonated molecule, a major product ion(s), isotopic matching, and chromatographic retention-time matching.

REFERENCES

1. Wylie, P.L., Agilent Technologies Application Note for GC/MS screening of pesticides, 2005. PN 5989-5076 EN
2. Ferrer, I.; Thurman, E.M., *Liquid Chromatography/Mass Spectrometry/Mass Spectrometry and Time-of-Flight Mass Spectrometry for the Analysis of Emerging Contaminants*, 2003, American Chemical Society Symposium Volume 850.
3. Ferrer, I.; Thurman, E.M., Analysis liquid chromatography/time-of-flight/mass spectrometry (LC/TOF/MS) for the analysis of emerging contaminants, *Trends Anal. Chem.*, **2003**, 22, 750–756.
4. Baumann, C.; Cintora, M.A.; Eichler, M.; Lifante, E.; Cooke, M.; Przyborowska, A.; Halket, J.M., A library of atmospheric pressure ionization daughter ion mass spectra based on wideband excitation in an ion trap mass spectrometer, *Rapid Commun. Mass Spectrom.*, **2000**, 14, 349–356.
5. Josephs, J.L.; Sanders, M., Creation and comparison of MS/MS spectral libraries using quadrupole ion trap and triple-quadruople mass spectrometers, *Rapid Commun. Mass Spectrom.*, **2004**, 18, 743–759.
6. Gergov, M.; Weinmann, W.; Meriluoto, J.; Uusitalo, J.; Ojanpera, L., Comparison of product ion spectra obtained by liquid chromatography/triple-quadrupole mass spectrometry for library search, *Rapid Commun. Mass Spectrom.*, **2004**, 18, 1039–1046.
7. Mueller, C.A.; Weinmann, W.; Dresen, S.; Scheiber, A.; Gergov, M., Development of a multi-target screening analysis for 301 drugs using a QTrap liquid chromatography/tandem mass spectrometry system and automated library searching, *Rapid Commun. Mass Spectrom.*, **2005**, 19, 1332–1338.
8. Ferrer, I.; García-Reyes, J.F.; Mezcua, M.; Thurman, E.M.; Fernández-Alba, A.R., Multi-residue pesticide analysis in fruit and vegetables by liquid chromatography-time-of-flight-mass spectrometry, *J. Chromatogr. A*, **2005**, 1082, 81–90.
9. Ferrer, I.; Thurman, E.M., Measuring the mass of an electron by LC/TOF-MS: a study of "twin ions", *Anal. Chem.*, **2005**, 77, 3394–3400.
10. Thurman, E.M.; Ferrer, I.; Zweigenbaum, J.A., High resolution and accurate mass analysis of xenobiotics in food, *Anal. Chem.*, **2006**, 78, 6702–6708.
11. Thurman, E.M.; Ferrer, Imma; Fernandez-Alba, A.R., Matching unknown empirical formulas to chemical structure using LC/MS TOF accurate mass and database searching: example of unknown pesticides on tomato skin, *J. Chromatogr. A*, **2005**, 1067, 127–134.
12. Thurman, E.M.; Ferrer, Imma; Zweigenbaum, J.A.; Garcia-Reyes, J.F.; Woodman, Michael, Fernandez-Alba, A.R., Discovering metabolites of post harvest fungicides in citrus with liquid chromatography/time-of-flight mass spectrometry and ion trap tandem mass spectrometry, *J. Chromatogr. A*, **2005**, 1082, 71–80.
13. Bobeldijk, I.; Vissers, J.P.C.; Kearney, G.; Major, H.; van Leerdam, J.A., Screening and identification of unknown contaminants in water with liquid chromatography and quadrupole-orthogonal acceleration-time-of-flight tandem mass spectrometry, *J. Chromatogr. A*, **2001**, 929, 63–74.
14. Laks, S.; Pelander, A.; Vuori, E.; Ali-Toppa, E.; Sippola, E.; Ojanpera, I., Analysis of street drugs in seized material without primary reference standards, *Anal. Chem.*, **2004**, 76, 7375–7379.
15. Pelander, A.; Ojanpera, I.; Laks, S.; Rasanen, I.; Vuori, E., Toxicological screening with formula-based metabolite identification by liquid chromatography/time-of-flight mass spectrometry, *Anal. Chem.*, **2003**, 75, 5710–5718.

16. Stolker, A.A.M.; Dijkman, E.; Niesing, W.; Hogendoorn, E.A., Identification of residues by LC/MS/MS according to the new European Union Guidelines: application to the trace analysis of veterinary drugs and contaminants in biological and environmental matrices, Chapter 3, pp. 32–51, in *Liquid Chromatography/Mass Spectrometry/Mass Spectrometry and Time-of-Flight Mass Spectrometry for the Analysis of Emerging Contaminants*, **2003**, American Chemical Society Symposium Volume 850.

17. Thurman, E.M.; Goolsby, D.A.; Meyer, M.T.; Mills, M.S.; Pomes, M.L.; Kolpin, D.W., A reconnaissance study of herbicides and their metabolites in surface-water of the Midwestern united-states using immunoassay and gas-chromatography mass-spectrometry, *Environ. Sci. Technol.*, **1992**, 26, 2440–2447.

CHAPTER 5

USE OF UPLC-(Q)TOF MS FOR RAPID SCREENING OF ORGANIC MICROPOLLUTANTS IN WATER

Félix Hernández, Juan V. Sancho, and María Ibáñez

Research Institute for Pesticides and Water, University Jaume I, Castellón, Spain

IN THIS CHAPTER, the potential of ultra-performance liquid chromatography (UPLC) coupled to (quadrupole) time-of-flight mass spectrometry ((Q)TOF MS) for screening of organic (micro) pollutants in water is illustrated. The combination of these two powerful techniques offers to the analyst an attractive and efficient way for rapid and wide-scope screening of a huge number of organic pollutants with very little sample manipulation. UPLC provides a fast chromatographic run with improved resolution, minimizing interferences from co-eluting peaks, and TOF MS allows performing full scan spectra with high sensitivity and mass accuracy. The applicability of this hyphenated technique for both target and non-target screening is discussed in this article. The most interesting applications are found in post-target screening, i.e., the selection of the compounds to be searched in the samples is made after MS data acquisition, where the use of narrow-window extracted ion chromatograms allows reducing the number of potential interferents making the screening more specific. Without the need of re-analyzing the samples, the presence of many compounds can be investigated months or even years later, taking profit of the abundant and useful information contained in TOF full spectra acquisitions. Regarding non-target screening, component detection algorithms and powerful chromatographic deconvolution software are required to manage the huge amount of MS data available. Experimental data are compared versus a user-built library (empirical and/or theoretical), which can contain hundreds/thousands of organic contaminants, facilitating a wide-scope, universal screening. When a compound is not found in the library, its deconvoluted accurate mass spectra can be used to propose its elemental composition. In this context, the availability of a QTOF MS instrument would allow performing MS/MS experiments with accurate mass measurements, facilitating the elucidation of non-target/unknown compounds as well as the safe confirmation of

Liquid Chromatography Time-of-Flight Mass Spectrometry: Principles, Tools, and Applications for Accurate Mass Analysis, Edited by Imma Ferrer and E. Michael Thurman
Copyright © 2009 John Wiley & Sons, Inc.

target compounds present in the samples. This strategy has been applied to different types of water samples (urban waste water, surface and ground water) and has allowed the detection of several pesticides, like terbutryn, simazine, and thiabenda-zole, among others, at low ppb levels. Antibiotics, such as ofloxacin or ciprofloxacin, and drugs of abuse, such as benzoylecgonine (a cocaine metabolite), have also been detected. The home-made theoretical library employed contained more than 500 compounds, including many pesticides and transformation products, antibiotics, and several drugs, although the number of entries could be easily increased to facilitate a wider screening.

5.1   INTRODUCTION

During the last decades, a large number of organic micro pollutants have been released into the environment as a result of anthropogenic activities. Environmental contaminants of recent concern are pharmaceuticals, pesticides (especially their transformation products), estrogens and other endocrine disrupting chemicals, such as degradation products of surfactants, algal and cyanobacterial toxins, disinfection by-products, and metalloids [1]. One of the main sources of these microcontaminants is untreated urban wastewaters and wastewater treatment plant (WWTP) effluents. Most current WWTPs are not designed to treat these types of substances and a high portion of emerging contaminants can escape elimination and enter the aquatic environment via sewage effluents [2]. Due to the recent world-wide concern about the consequences of this contamination, water has become the subject of much discussion, being its analysis of particularly high priority.

However, the analytical determination of these micropollutants in water is not an easy task: firstly, because of the high number and wide variety of analytes that can potentially reach the water; secondly, due to the complexity of some matrices, such as wastewater or leachate samples; and, finally, because of the usually extremely low concentration levels in which they are present. These low levels make, in general, necessary the application of preconcentration procedures in order to achieve the required sensitivity. Liquid-liquid extraction (LLE) or solid-phase extraction (SPE) have been the most widely used methods in sample preconcentration for dif-ferent compound classes, like pesticides, pharmaceuticals, or estrogens. However, nowadays, for aqueous samples, LLE has been almost completely replaced by SPE. Using SPE, smaller sample volumes are needed and a large number of polar com-pounds can be efficiently extracted with little solvent consumption. The possibility of using on-line SPE, especially in SPE-LC/MS based methods, is another important advantage, because of the easy automation of the system, and the excellent detection limits achieved.

Under these circumstances, development of new and more sensitive analytical methods for the detection of chemicals is necessary. Due to the huge number of potential organic pollutants that could be present in the environment, wide-scope screening methods are preferred in order to have a general overview on sample composition. Screening methods should be able to (quickly) detect the presence of as many contaminants as possible in a sample, preferably with little sample manipulation. In general, two alternatives can be considered, depending on the objective of the screening and, especially, on the instrumentation available:

A. Target screening methods, where the analytes are preselected before the analysis (pre-target screening) or after data acquisition (post-target screening). In some cases, they can also give a semi-quantitative estimation of the concentration level, but, in general, a second analysis is required to correctly quantify the analytes detected and/or confirm their identity

B. Non-target screening methods, where all compounds eluting from the chromatographic analytical column should be detected and identified without any kind of previous selection. With the obvious limitations derived from the chromatographic and ionization requirements, the ideal option would be to investigate all unknown compounds present in a sample. At present, the identification of non-target compounds in environmental samples is still a challenge for analytical chemists due to the inherent major difficulties of this subject.

For rapid laboratory responses in cases of emergency and to have as much as information as possible on environmental water samples analyzed, it would be ideal to have at one's disposal a methodology for screening the widest number of compounds, which also allowed a reliable identification, in only one analysis.

Until the beginning of the 1990s, non-polar hazardous compounds were the focus of interest and awareness as priority pollutants, thus the analysis of environmental samples was traditionally dominated by the use of gas chromatography–mass spectrometry (GC-MS). However, because most of organic contaminants, and even more their metabolites and transformation products (TPs), are highly polar and water-soluble, liquid chromatography–mass spectrometry (LC-MS) has become the method of choice for their determination.

Since the introduction of atmospheric pressure ionization techniques, LC-MS, and lately LC-MS/MS, have played an increasingly important role in environmental analysis. This is proved by, for example, the growing number of papers published in the last years concerning the determination of organic pollutants in water by LC-MS(/MS) [3, 4]. Electrospray (ESI) and atmospheric pressure chemical ionization (APCI) are applicable for the analysis of a broad range of compounds, including polar species (and even ionic), non-volatile, and thermally labile, without need of derivatization. Moreover, ESI and APCI provide high sensitivity, which is mandatory for environmental analysis where contaminants are often found at ng/L levels. Another important advantage of LC-MS in the analysis of water samples deals with the possibility of performing direct aqueous sample injection, without the need of an extraction step, which is nowadays feasible in many cases due to the excellent sensitivity reached by modern LC-MS equipments. When necessary, a preconcentration can be carried out in order to achieve the required sensitivity, with SPE (off- and on-line modes) being the most widely applied technique. One of the main drawbacks is, at present, the non-availability of mass spectral libraries for LC-MS. Unfortunately, most libraries with appropriate sets of data were created for electron impact ionization data under standardized conditions. For the soft-ionization techniques, like ESI and APCI, fragment patterns are obtained by collision-induced fragmentation in the electrospray-transport region, in quadrupole collision cells or in an ion trap, and are strongly dependent on operational and instrumental parameters. Therefore, there are no standard conditions, and user libraries are usually generated at best for an individual instrument [5].

With recent advances in mass spectrometry, the orthogonal-accelerated time of flight (oa-TOF) mass spectrometer has become available. The benefits of using TOF analyzer relays on its measuring principle, that allows to perform full scan acquisitions with superior sensitivity and high mass accuracy (2–5 ppm). These characteristics together with its higher mass resolving power (>10,000 FWHM) are very attractive when developing analytical methodology for screening, confirmation, and elucidation of organic pollutants at relevant environmental levels.

Even more useful in terms of confirmatory analysis is the hybrid analyzer quadrupole-TOF (QTOF). The development of this analyzer has offered to the analyst a powerful tool for the determination of pollutants in the environment. QTOF presents all the advantages indicated above for TOF, as it can be used in TOF-mode. Besides, it is able to obtain a full product ion scan with accurate mass. Thus, while the measurement of accurate mass in TOF allows setting up the elemental composition of a compound, QTOF also allows establishing the elemental composition of all product ions obtained, which is very helpful in the elucidation process of unknowns. The accurate mass together with the acquisition of the full scan spectra of product ions is also a powerful tool for the unequivocal confirmation of positives (target analytes). However, and despite the indubitable potential of QTOF, its use in the environmental field is still very limited, mainly due to its high cost.

Regarding LC, due to the very high selectivity of tandem MS detection, the importance of good chromatographic separation is often neglected. However, a poor chromatographic separation may lead to ion suppression and isobaric interferences, especially in trace analysis, where failure to completely separate compounds from each other and from the matrix components may result in the non-detection of low concentration compounds [6]. A novel approach to chromatographic separation is ultra performance liquid chromatography (UPLC). This new technology provides a fast, high resolution separation, which increases LC/MS sensitivity and minimizes matrix interference arising from minimal sample preparation. However, to fully take profit of UPLC advantages, modern mass analyzers with rapid spectra acquisition are required.

When coupled, these new powerful technologies provide a potent analysis, very rich in information on sample composition. Because of the recent development of both technologies, very few applications using UPLC-(Q)TOF MS have been reported in the environmental field [6, 7]. Until now, most of the reported applications are found in the metabolite profiling field [8–11], impurity profiling of pharmaceutical drug substances [12], metabonomics [13], or veterinary drugs in urine [14]. All these references have been published in the last three years, showing that this is a "hot topic" that surely will be widely investigated in the near future due to the excellent analytical characteristics and new expectations derived from this powerful hyphenated technique.

5.2   TARGET ANALYSIS

As it has been commented before, when dealing with environmental analysis, screening methods are normally developed with the aim of rapidly detect the presence of as many contaminants as possible in a sample. In LC-MS target screening methods two alternatives might be considered depending on the instrument available:

A. **Pre-target screening**, where the analytes are preselected before the analysis and therefore other positives cannot be revealed. Typically these methods are applied to a list of priority pollutants that rarely exceeds 100 compounds. As this approach is focused on a few analytes, it is normally feasible to have a satisfactory quantification without need of re-injecting the sample

B. **Post-target screening**, where all the compounds eluting from the chromatographic column are measured by MS and selected m/z ions are extracted afterwards from the total ion current chromatogram. A lot of useful information on sample composition can be obtained in this way, but a second analysis is required to report the concentration levels of the compounds detected.

The use of selected reaction monitoring (SRM) mode with triple quadrupole analyzers (QqQ), makes these instruments ideal for the pre-target screening, due to their excellent characteristics of sensitivity, selectivity, little sample manipulation, and rapid analysis. However, the low sensitivity in full scan mode and their low mass resolution make unadvisable their use in post- and non-target screening of organic pollutants.

Regarding ion traps (IT) analyzers, the possibility of performing full spectra acquisitions is attractive for screening purposes and they might be used for post-target and non-target applications. However, these analyzers generate nominal mass data, being less adequate for confirmation or elucidation purposes. Another drawback of IT instruments comes from the difficulties of measuring product ions with m/z lower than 30% of that of precursor ion, which may limit their applicability in some particular cases.

On the contrary, the ability of TOF-MS analyzers to provide full scan spectra with high sensitivity and mass accuracy makes them an interesting choice for post-target and non-target screening. These characteristics allow detecting a huge number of potential contaminants in a sample without re-analysis, with the obvious limitations derived from requirements of LC-MS in relation to the chromatographic, ionization, and fragmentation processes. This high level of multiresiduality cannot be easily achieved by quadrupole mass analyzers (both Q and QqQ, working in selected ion monitoring (SIM) or SRM mode, respectively) due to the need to pre-define the masses to be monitored and to the difficulty to reduce the dwell time below a threshold value maintaining a suitable sensitivity.

An illustrative example of post-target screening using LC-TOF MS has been reported by our own research group [15] when analyzing an urban wastewater sample collected from an agricultural area with predominance of citric crops. The detection of the post-harvest fungicide imazalil led us to suspect the possible

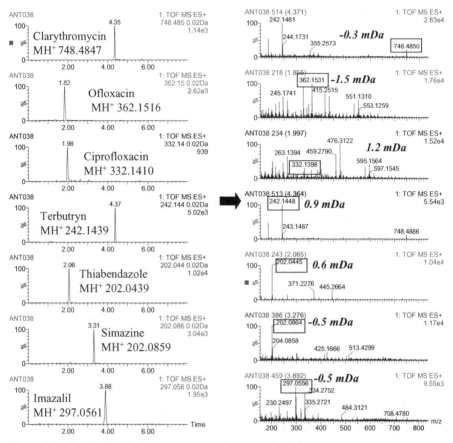

Figure 5.1. LC-TOF MS chromatograms and spectra of a influent wastewater treatment plant sample containing the antibiotics clarythromycin, ofloxacin, and ciprofloxacin, and the pesticides terbutryn, thiabendazole, simazine, and imazalil.

presence of other fungicides. After extracting chromatograms at the specific masses of several fungicides, the presence of thiabendazole ($[M + H]^+$ m/z 202.0439) was also confirmed.

Another example is shown in Figure 5.1. After analyzing an influent WWTP sample, the antibiotics ofloxacin, ciprofloxacin, and clarythromycin were detected. As the WWTP was located in an area with an important agricultural activity, the presence of several pesticides was checked in a post-target way. This led us to the detection of two herbicides (terbutryn and simazine) and two fungicides (thiabendazole and imazalil) widely used in the area.

Moreover, the elevated mass resolution of TOF analyzers allows narrowing the mass window when extracting a specific mass from the full scan dataset, which leads to a substantial reduction of the chemical noise and facilitate the detection of the screened compound in the eXtracted Ion Chromatogram (XIC, also named EIC or RIC depending on the manufacturer). The benefits of reducing the mass window from 1 Da scale (similar to quadrupole or ion trap analyzers) down to 10–20 mDa

Oxolinic acid, $C_{14}H_{12}NO_3F$, $(M+H)^+$ 262.0715

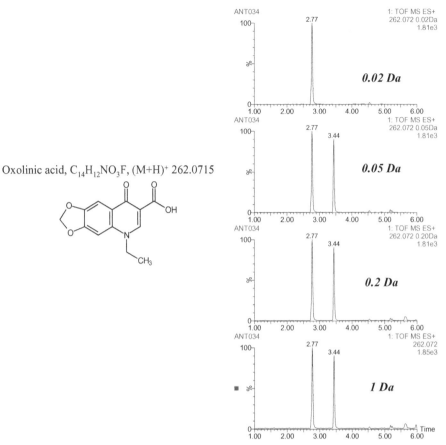

Figure 5.2.   Comparison of different mass extraction windows (1, 0.2, 0.05, and 0.02 Da) for *m/z* 262.0715 in the LC-TOF MS analysis of a hospital wastewater sample, where the antimicrobial oxolinic acid was detected.

have been reported in the bibliography [6, 7, 16, 17]. As an example, Petrovic et al. [6] noticed a 15-fold increase of the signal-to-noise ratio in the analysis of pharmaceuticals in wastewater, by reducing the mass window from 100 to 20 mDa. This reduction of the mass window also resulted in an almost complete loss of the interferences from the isobaric contaminant ions. Decreasing mass window leads to an improved selectivity as Figure 5.2 shows, where XICs corresponding to *m/z* 262.0715 (oxolinic acid) using different mass extraction windows are depicted for a raw hospital wastewater. As it can be seen, when a mass window of 1 Da was selected, two chromatographic peaks corresponding to the antibiotic oxonilic acid and to an isobaric compound were observed. After decreasing the mass window down to 0.2 Da or 0.05 Da still the two peaks were observed. Finally, selecting a mass window of 0.02 Da led to only one peak, which corresponded to oxolinic acid.

When using narrower mass windows (microwindow XIC, mw-XIC), the reliability of the mass accuracy attainable by the TOF analyzer is of outstanding importance. Besides, a drawback of using mw-XIC comes from the significant mass errors that may be produced by coeluting isobaric interferents that cannot be resolved by the analyzer [17]. The extent of mass errors might be so high so that the screened compound might fall out of the monitored mass window, which would lead to reporting a false negative. An illustrative example was found by Benotti et al. [18] when screening pharmaceuticals in wastewater effluents. High mass errors (20 mDa) were observed for caffeine due to presence of the ^{13}C isotope peak from a co-eluting compound with a mass 1 Da lower than the analyte. In the case that XICs had been reconstructed with smaller mass window (±5–10 mDa), the presence of caffeine would have been masked, and a false negative would have been reported. Thus, when performing screening in real samples one should be cautious and avoid using unreasonably narrow mass windows. As a compromise between improving baseline noise and signal-to-noise ratio and preventing reporting false negatives, a 20–50 mDa mass window would be recommendable.

The use of QTOF in MS/MS mode implies the preselection of the analytes, as it is necessary to know the analyte m/z in order to filter it in the quadrupole. Obviously, to avoid preselecting the analytes, the QTOF instrument might be used in TOF mode, but the genuine characteristics of this hybrid analyzer when working in MS/MS mode would not be revealed. As a consequence, the great potential of QTOF for screening purposes is actually found in pre-target applications, although used as a TOF instrument it would be also appropriate for post- and non-target screening. Its main advantage comes from the acquisition of the complete product ion spectra with accurate mass, allowing the simultaneous screening and reliable confirmation of the selected analyte in one injection.

We have tested the applicability of UPLC-QTOF for the confirmation of antibiotics previously detected with a triple quadrupole in environmental waters. The acquisition of the full product ion spectra at accurate mass allowed us the unequivocal confirmation of the compounds detected. As an example, Figure 5.3 shows the QTOF MS chromatograms and product ion spectra for the three antibiotics detected in the hospital wastewater sample commented before. Table 5.1 shows the mass errors obtained for the parent ion and the product ions of the selected compounds. As it can be seen, mass errors were in general lower than 1 mDa, being in all cases below 2 mDa, which gave a high confidence in the identification of the compounds detected.

Regarding the screening by QTOF instruments, there has been a drawback related with the limited number of compounds able to be monitored simultaneously. The acquisition in a sequential mode by the quadrupole and the relatively high time needed by the early TOF instruments to have an adequate response for the measured product ion spectrum (around 1 second) notably reduced the number of analytes that could be selected in a time fraction. For these reasons, the use of QTOF for screening purposes had been normally limited to a few pre-target compounds. For example, LC-QTOF was tested in the multiresidue screening of around 30 pesticides and transformation products, and it was considered a valuable tool for confirmative analysis [19]. Another example of using QTOF for screening purposes can be found in the determination of pharmaceuticals in water [20]. In this case, a multi-residue method including 13 pharmaceuticals was developed for the simultaneous screening

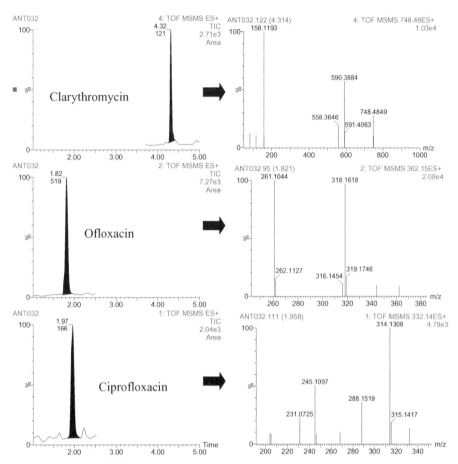

Figure 5.3.   LC-QTOF MS chromatograms and product ion spectra of a hospital wastewater sample containing clarythromycin, ofloxacin, and ciprofloxacin (for more details see Table 5.1).

and confirmation at the low µg/L level. Recent QTOF instruments, with faster spectra acquisition, have allowed a notable increase in the number of screened compounds.

Another possibility to increase the number of analytes monitored is the use of automated MS to MS/MS switching as demonstrated by Bobeldijk et al. [21]. This approach is based on the possibility to automatically change from MS to MS/MS mode when the compound of interest is eluting from the analytical column. The instrument is initially set up as a TOF acquiring in full scan mode; when a specific mass exceeds a predefined number of counts, the instrument automatically changes to MS/MS mode recording the product ion spectrum at this mass, returning to TOF mode once the spectrum is acquired. This approach was tested for six pesticides, used as model compounds, showing its suitability for screening and identification. In this way, it is not necessary to preselect the analytes before the screening. The

TABLE 5.1. Exact Mass Measurements: Fragment Mass Errors for Several Antibiotics Found in Hospital Wastewater Using UPLC-QTOF MS

Compound	Empirical Formula[a]	Exact Mass[b]	Experimental Accurate Mass	Mass Error (mDa)
Clarythromycin	$C_{38}H_{70}NO_{13}$	748.4847	748.4849	0.2
	$C_{30}H_{56}NO_{10}$	590.3904	590.3884	2
	$C_{29}H_{52}NO_9$	558.3642	558.3646	0.4
	$C_8H_{16}NO_2$	158.1181	158.1193	1.2
	$C_6H_{14}NO$	116.1075	116.1071	−0.4
	C_5H_7O	83.0497	83.0490	−0.7
Ofloxacin	$C_{18}H_{21}N_3O_4F$	362.1516	36.21517	0.1
	$C_{18}H_{19}N_3O_3F$	344.1410	344.1425	1.5
	$C_{17}H_{21}N_3O_2F$	318.1618	318.1618	0
	$C_{11}H_{10}N_2O_2F$	261.1039	261.1044	0.5
	C_4H_8N	70.0657	70.0652	−0.5
Ciprofloxacin	$C_{17}H_{18}N_3O_3F$	332.1410	332.1405	−0.5
	$C_{17}H_{16}N_3O_2F$	314.1305	314.1308	0.3
	$C_{16}H_{18}N_3OF$	288.1512	288.1519	0.7
	$C_{14}H_{14}N_2OF$	245.1090	245.1097	0.7
	$C_{11}H_9N_2OF$	204.0699	204.0688	−1.1

(a) top: $[M + H]^+$ ion; down: fragment ions.
(b) There is a difference of around 0.5 mDa between the theoretical mass and the calculated one, as the software used adds up or subtracts the mass of a hydrogen atom instead of the mass of a proton [28].

most noticeable drawback is found in the need to predefine a threshold value above of which a compound is considered relevant.

5.3 NON-TARGET ANALYSIS

Target-compound monitoring is often insufficient to assess the quality of environmental waters as only a limited number of analytes is recorded; so other potentially harmful contaminants, unexpected or not included in routine monitoring protocols, which may be present in the samples would not have been reported. In the case of GC-MS amenable analytes, identification of non-target compounds is sometimes feasible [22, 23] as the use of forward-search methods enables searching the large National Institute for Standards and Testing (NIST) library, among others. Unluckily, in the case of LC-MS/MS, identification of unknowns is a complex and time consuming process [15, 21, 24], the success of which depends, in some way, on the availability of databases or libraries where the search of an elucidated elemental composition can be performed [15, 24, 25]. Due to the difficulties associated to this type of analysis, rather than performing a general screening for unknown compounds, the identification effort should be focused on relevant compounds and not on all unknown peaks present in the sample, as obviously not all LC-MS amenable compounds present in environmental waters are necessarily relevant for man or environment [24]. For example, Ibáñez et al. [7] focused the screening of water samples on pesticides, although other emerging contaminants, like antibiotics or some pharmaceuticals widely used, were also investigated.

As stated before, single quadrupole and triple quadrupole instruments are not adequate for non-target analysis as they do not typically operate in full scan mode because of the lack of sensitivity. Regarding ion traps, although MS/MS libraries for this instrument have been produced, they have not been useful due to the irre-producibility of fragmentation [26]. In the case of hybrid Q-trap with a linear ion trap there are also some limitations on the number of ions that may be selected because of dwell times [26]. However, a TOF MS analyzer can provide a possible solution for this problem as it combines high full-spectral sensitivity with high resolution mass spectra allowing any ionizable component in a sample to be accurate mass-measured. TOF MS allows the user to obtain a notable amount of chemical information on sample composition in a single experiment making this technique very attractive to search for analytes in a post-target way or to perform a non-target analysis. This last aspect will be discussed in more detail in the following paragraphs.

When screening a sample for unexpected compounds, it becomes difficult to pick out individual ions, especially when the matrix is complex or the concentration of the analyte is low. Under these circumstances, it is necessary the application of powerful deconvolution softwares produce pure spectra for each individual compo-nent. The process can be divided in two phases. The first step is the chromatographic peak deconvolution. Based on parameters that describe the chromatographic peaks, this software must be able to discriminate real compounds present in the sample from background ions and create a list of components. On a second step, the software must be able to perform a library search to match the proposed components to the existing entries in the library. However, there is still a limitation in the software developed until now as the use of exact-mass is not included in the database search for candidate assignation.

It is interesting to comment on the methodology applied when using the soft-ware ChromaLynx (from Waters), as the authors have some experience on it. In this case, after performing a user-built library search (in nominal mass), an automatic accurate mass confirmation is carried out. The formula from the library hit is then submitted to an elemental composition calculator included in the software package and the most intense ions are confirmed or rejected by accurate mass criteria. In general, when working with LC-MS, only the (de)protonated molecule, a favorable isotopic pattern and/or an in-source fragment ion (if exist) are typically observed. Figure 5.4 shows the information obtained in non-target screening when using ChromaLynx software. In this case, the finding of benzoylecgonine, a metabolite of cocaine, in a wastewater effluent has been selected as example. At the top (Figure 5.4A), the positive total ion chromatogram (TIC) is displayed, with bottom triangles to signify where a peak has been found and with selected benzoylecgonine marked as a top triangle. Below this section (Figure 5.4B), a list with all detected compo-nents is shown. In Figure 5.4C, the TIC is displayed again, but only the components identified in the library with a match higher than 700 are marked. The results on the left (Figure 5.4D) show the candidate list, i.e., those components whose spectrum

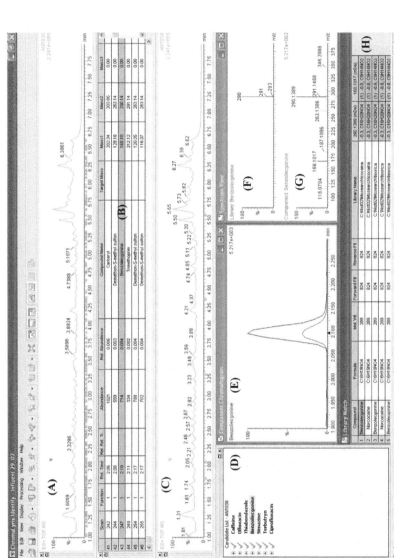

Figure 5.4.   ChromaLynx browser with accurate mass confirmation for a finding of benzoylecgonine, a metabolite of cocaine, in the effluent of a WWTP. (A) Total ion chromatogram with all identified compounds (marked as bottom triangles) and with selected benzoylecgonine (marked as a top triangle). (B) List with all detected components. (C) Total ion chromatogram with identified compounds with library match higher than 700 (marked as bottom triangles) and with selected benzoylecgonine (marked as a top triangle). (D) Candidate list wit a library match higher than 700. (E) Overlap of the two main ions of benzoylecgonine used for deconvolution. (F) Library mass spectrum of benzoylecgonine at nominal masses. (G) Deconvoluted mass spectrum of benzoylecgonine from water extract. (H) Library match and accurate

has been matched with one or more library entries with a match higher than 700. Figure 5.4E shows the component chromatograms, where the two deconvoluted ions are overlapped. On its right, the theoretical library mass spectrum of benzoylecgonine (Figure 5.4F) and the deconvoluted mass spectrum of this compound from water (Figure 5.4G) are shown. Finally, Figure 5.4H shows the library match and the automatic accurate mass confirmation of the two deconvoluted ions. As it can be seen, in this case, mass errors were lower than 1 mDa for both the parent (m/z 290) and the in-source fragment (m/z 168).

Due to the non-availability of LC-MS mass spectral libraries, user in-house libraries are normally generated. Two possibilities are plausible: to build an empirical "home-made" library of spectra by injecting standard solutions into the system in full acquisition mode; or, to create a theoretical "computer-made" library, which includes a theoretical spectrum of each compound. Although a theoretical library allows including a high number of compounds without the need of injecting standards [7], its main limitation is the absence of information on in-source fragments, which in some cases can help in the reliable confirmation of the compound's identity.

Thurman et al. [27] created a homemade Microsoft Access database of 350 pesticides amenable to positive ion electrospray containing their exact monoisotopic mass. This database presented the limitation that could only be searched in manual mode by entering the accurate mass of the detected compound, which had to be also found in a manual mode. Thus, automation of the database was a problem that needed to be addressed for satisfactory routine analysis. The same group [26] used a semi-automated "molecular-feature" and database searching consisting of the exact monoisotopic mass of 100 compounds, one exact mass product ion for each compound and retention time, to identify pesticides in food and water samples. The software used extracted a list of accurate mass ions, which were compared to the monoisotopic exact masses of the database. This methodology notably improved the searching of compounds in samples, although there were still some limitations in this approach, as it was not feasible to perform an automated isotopic distribution matching. More recently, Ibáñez et al. [7] have created a "home-made" empirical library and a "computer-made" library, containing more than 500 organic contaminants, including pesticides, pesticide transformation products, antibiotics, some widely used pharmaceuticals, and some drugs of abuse. UPLC-TOF MS in combination with ChromaLynx software was used in this work for analysis of non-targeted compounds in 12 water samples of different types. Concretely, five surface water samples, three influent, and three effluent WWTP samples, as well as a raw hospital wastewater were analyzed. The most commonly detected pesticides were the postharvest fungicides imazalil and thiabendazole, the triazine herbicides terbuthylazine, simazine and terbutryn, and the phenylurea herbicide diuron. Antibiotics like ofloxacin and ciprofloxacin, and the pharmaceutical paracetamol were also detected. Other contaminants identified in the wastewater samples were caffeine and benzoylecgonine, a cocaine metabolite. It is noteworthy that in almost all cases, confirmation of the detected compound was favored by the presence of isotopic peaks or in-source fragments. Table 5.2 shows data obtained in the analysis of a influent WWTP sample.

TABLE 5.2.   Summary of Non-target Compounds Identified in an Influent WWTP Sample, Collected in July 2005

Compound	Empirical Formula[a]	Exact Mass[b]	Experimental Accurate Mass	Mass Error (mDa)	Reference for Fragmentation[c]
Simazine	$C_7H_{13}N_5{}^{35}Cl$	202.0859	202.0864	0.6	[29]
	$C_7H_{13}N_5{}^{37}Cl$	202.0830	204.0848	1.8	
	$C_5H_9N_5Cl$	174.0546	174.0540	−0.6	
Terbutryn	$C_{10}H_{20}N_5S$	242.1439	242.1448	0.9	[29]
	$C_6H_{12}N_5S$	186.0813	186.0816	0.9	
Thiabendazole	$C_{10}H_8N_3S$	202.0439	202.0441	0.2	[16]
	$C_9H_7N_2S$	175.0330	175.0335	0.5	
Ofloxacin	$C_{18}H_{21}N_3O_4F$	362.1516	362.1530	1.4	[30]
	$C_{17}H_{21}N_3O_2F$	318.1619	318.1623	0.4	
Ciprofloxacin	$C_{17}H_{19}N_3O_3F$	332.1410	332.1408	−0.2	[30]
	$C_{16}H_{19}N_3OF$	288.1512	288.1515	0.3	
Caffeine	$C_8H_{11}N_4O_2$	195.0882	195.0899	1.7	[31, 32]
	$C_6H_8N_3O$	138.0667	138.0668	0.1	
Benzoylecgonine	$C_{16}H_{20}NO_4$	290.1392	290.1389	−0.3	[33]
	$C_9H_{14}NO_2$	168.1025	168.1017	−0.8	

Empirical formulae, as well as experimental accurate masses and mass errors are shown.
(a) top: $[M + H]^+$ ion; down: isotopic or in-source fragments.
(b) There is a difference of around 0.5 mDa between the theoretical mass and the calculated one, as the software used adds up or subtracts the mass of a hydrogen atom instead of the mass of a proton. [28]
(c) Fragmentation data have been deduced from cited reference.

In this type of analysis, the use of UPLC is of great importance as the higher resolution separation provided by this system allows minimizing interferences from co-eluting peaks, which favors an adequate component deconvolution. Moreover, UPLC provides short chromatographic runs (around 8 min) versus the typical around 30-min runs when using HPLC.

Once the detected compound has been assigned to a possible structure, an unequivocal confirmation could be carried out with a QTOF instrument, by acquiring the full product ion spectra with accurate mass. When confirming positive findings by QTOF, both the exact mass and the relative intensity of all available product ions of a sample could be compared with those of the reference standard, if available. Under these circumstances, the confirmation achieved by QTOF can be considered as the ultimate confirmation of the analyte identity [15]. Another important advantage of QTOF in confirmatory applications is the capability to obtain abundant fragmentation without any significant interference. The use of QTOF minimizes the limitations of TOF instruments when working with in-source fragment ions, as the selection of a precursor ion in the first quadrupole increases the confidence about the product ion origin and decreases the chemical noise. Additionally, the low chemical noise and/or the efficient fragmentation produced in the collision cell increase the number and relative intensity of product ions when using QTOF, allowing to improve the confirmation quality and also to confirm positive samples at concentration levels close to the limit of detection.

On the other hand, if the detected compound is not included in the library, the high mass accuracy provided by TOF and any potential in-source fragmentation generated, can be used to try to elucidate its elemental composition. Additionally, the possibility of applying different filtering technology based on isotope ratio deviations, may aid to simplify the hit list. But, undoubtedly, the valuable data given by the product ion spectra with accurate mass when using a QTOF provide relevant structural information, helping the analyst to elucidate the structure of an unknown compound.

5.4 CONCLUSIONS

Hyphenation of UPLC and (Q)TOF MS is an efficient and advanced approach for the rapid screening of organic (micro) pollutants in water. On one hand, the UPLC system provides a fast chromatographic run with improved resolution, allowing to minimize interferences from co-eluting peaks. On the other hand, the inherent characteristics of QTOF MS make this analyzer very attractive for the determination of micro-contaminants in water samples.

The UPLC-TOF potential for screening purposes comes from its ability to acquire full scan spectra with high sensitivity, which increases the multiresiduality ability and facilitates its application to the monitoring of a high number of compounds that might be present in a water sample. Additionally, the use of mw-XIC also allows reducing the number of interferents making the screening more specific. To manage the huge amount of MS data available, powerful deconvolution softwares are required.

The potential of QTOF MS to perform MS/MS experiments with accurate mass measurements allows a safe confirmation of the compounds present in the samples and facilitates the confident identification of non-target/unknown compounds. However, using a hybrid QTOF for screening purposes normally limits the number of analytes included in the method due to the need of preselecting the ion parent to be isolated in the quadrupole.

UPLC-TOF MS will surely become a powerful analytical tool for screening and identification of organic pollutants in the near future. To facilitate this task, home-made theoretical and/or empirical libraries are, at the moment, necessary. They should contain as many contaminants as possible for wide-scope, universal screening. Libraries should include the use of exact-mass in the algorithm search, as well as fragment ions when using empirical libraries, for an efficient and reliable candidate assignation.

The use of hybrid QTOF instead of a TOF analyzer would also allow the access to the new MS^E acquisition mode. In this mode, low and high collision energy full-scan acquisitions can be performed simultaneously, which results in valuable fragmentation information to be used in the confirmation and elucidation processes of unknowns. Besides, the possibility of performing additional MS/MS experiments gives the researcher the product ion accurate mass spectra which are much useful for elucidation purposes.

REFERENCES

1. Zwiener, C.; Frimmel, F.H., LC-MS analysis in the aquatic environment and water treatment technology—a critical review. Part II: Applications for emerging contaminants and related pollutants, microorganisms and humic acids, *Anal. Bioanal. Chem.*, **2004**, 378, 862–874.

2. Petrovic, M.; Gonzalez, S.; Barceló, D., Analysis and removal of emerging contaminants in wastewater and drinking water, *Trends Anal. Chem.*, **2003**, 22, 685–696.

3. Richardson, S.D., Water analysis: emerging contaminants and current issues, *Anal. Chem.*, **2007**, 79, 4295–4324.

4. Richardson, S.D., Water analysis: emerging contaminants and current issues, *Anal. Chem.*, **2005**, 77, 3807–3838.

5. Zwiener, C.; Frimmel, F.H., LC-MS analysis in the aquatic environment and in water treatment—a critical review. Part I: Instrumentation and general aspects of analysis and detection, *Anal. Bioanal. Chem.*, **2004**, 378, 851–861.

6. Petrovic, M.; Gros, M.; Barceló, D., Multi-residue analysis of pharmaceuticals in wastewater by ultra-performance liquid chromatography-quadrupole-time-of-flight mass spectrometry, *J. Chromatogr. A*, **2006**, 1124, 68–81.

7. Ibáñez, M.; Sancho, J.V.; McMillan, D.; Rao, R.; Hernández, F., Rapid non-target screening of organic pollutants in water samples by ultra performance liquid chromatography-time of flight mass spectrometry, *Trends Anal. Chem.*, **2008**, 27, 481–489.

8. Castro-Pérez, J.; Plumb, R.; Granger, J.H.; Beattie, I.; Joncour, K.; Wright, A., Increasing throughput and information content for in vitro drug metabolism experiments using ultra-performance liquid chromatography coupled to a quadrupole time-of-flight mass spectrometer, *Rapid Commun. Mass Spectrom.*, **2005**, 19, 843–848.

9. Kawanishi, H.; Toyo'oka, T.; Ito, K.; Maeda, M.; Hamada, T.; Fukushima, T.; Kato, M.; Inagaki, S., Rapid determination of histamine and its metabolites in mice hair by ultra-performance liquid chromatography with time-of-flight mass spectrometry, *J. Chromatogr. A*, **2006**, 1132, 148–156.

10. Johnson, K.A.; Plumb, R.J., Investigating the human metabolism of acetaminophen using UPLC and exact mass oa-TOF MS, *J. Pharmac. Biomed. Anal.*, **2005**, 39, 805–817.

11. Zhao, X.; Wang, W.; Wang, J.; Yang, J.; Xu, G., Urinary profiling investigation of metabolites with *cis*-diol structure from cancer patients based on UPLC-MS and HPLC-MS as well as multivariate statistical analysis, *J. Sep. Sci.*, **2006**, 29, 2444–2453.

12. Jones, M.D.; Plumb, R.S., The application of sub-2-μm particle liquid chromatography-operated high mobile linear velocities coupled to orthogonal accelerated time-of-flight mass spectrometry for the analysis of ranitidine and its impurities, *J. Sep. Sci.*, **2006**, 29, 2409–2420.

13. Yin, P.; Zhao, X.; Li, Q.; Wang, J.; Li, J.; Xu, G., Metabonomics Study of intestinal fistulas based on ultraperformance liquid chromatography coupled with Q-TOF mass spectrometry (UPLC/Q-TOF MS), *J. Proteomer. Res.*, **2006**, 5, 2135–2143.

14. Kaufmann, A.; Butcher, P.; Maden, K.; Widmer, M., Ultra-performance liquid chromatography coupled to time of flight mass spectrometry (UPLC-TOF): a novel tool for multiresidue screening of veterinary drugs in urine, *Anal. Chim. Acta*, **2007**, 586, 13–21.

15. Ibáñez, M.; Sancho, J.V.; Pozo, O.J.; Niessen, W.M.A.; Hernández, F., Use of quadrupole time-of-flight mass spectrometry in the elucidation of unknown compounds present in environmental water, *Rapid Commun. Mass Spectrom.*, **2005**, 19, 169–178.

16. Hogenboom, A.C.; Niessen, W.M.A.; Little, D.; Brinkman, U.A.Th., Accurate mass determination for the confirmation and identification of organic microcontaminants in surface water using on-line solid-phase extraction liquid chromatography electrospray orthogonal-acceleration time-of-flight mass spectrometry, *Rapid Commun. Mass Spectrom.*, **1999**, 13, 125–133.

17. Sancho, J.V.; Pozo, O.J.; Ibáñez, M.; Hernández, F., Potential of liquid chromatography/time-of-flight mass spectrometry for the determination of pesticides and transformation products in water, *Anal. Bioanal. Chem.*, **2006**, 386, 987–997.

18. Benotti, M.J.; Ferguson, P.L.; Rieger, R.A.; Iden, C.R.; Heine, C.E.; Brownawell, B.J., in *Liquid Chromatography/Mass Spectrometry, MS/MS and Time-of-Flight MS Analysis of Emerging Contaminants*, (eds Ferrer, I.; Thurman, E.M.) ACS Symposium series, American Chemical Society, Washington, DC, **2003**, 850, 109–127.

19. Hernández, F.; Pozo, O.J.; Sancho, J.V.; López, F.J.; Marín, J.M.; Ibáñez, M., Strategies for quantification and confirmation of multi-class polar pesticides and transformation products in water by LC-MS2 using triple quadrupole and hybrid quadrupole-time of flight analysers, *Trends Anal. Chem.*, **2005**, 24, 596–612.

20. Stolker, A.A.M.; Niesing, W.; Hogendoorn, E.A.; Versteegh, J.F.M.; Fuchs, R.; Brinkman, U.A.Th., Liquid chromatography with triple-quadrupole or quadrupole-time of flight mass spectrometry for screening and confirmation of residues of pharmaceuticals in water, *Anal. Bioanal. Chem.*, **2004**, 378, 955–963.

21. Bobeldijk, I.; Vissers, J.P.C.; Kearney, G.; Major, H.; van Leerdam, J.A., Screening and identification of unknown contaminants in water with liquid chromatography and quadrupole-orthogonal acceleration-time-of-flight tandem mass spectrometry, *J. Chromatogr. A*, **2001**, 929, 63–74.

22. Leandro, C.C.; Hancock, P.; Fussell, R.J.; Keely, B.J., Quantification and screening of pesticide residues in food by gas chromatography-exact mass time-of-flight mass spectrometry, *J. Chromatogr. A*, **2007**, 1166, 152–162.

23. Almeida, C.; Serôdio, P.; Florêncio, M.H.; Nogueira, J.M.F., New strategies to screen for endocrine-disrupting chemicals in the Portuguese marine environment utilizing large volume injection–capillary gas chromatography–mass spectrometry combined with retention time locking libraries, *Anal. Bioanal. Chem.*, **2007**, 387, 2569–2583.

24. Bobeldijk, I.; Stoks, P.G.M.; Vissers, J.P.C.; Emke, E.; van Leerdam, J.A.; Muilwijk, B.; Berbee, R.; Noij, Th.H.M., Use of quadrupole time-of-flight mass spectrometry in the elucidation of unknown compounds present in environmental water, *J. Chromatogr. A*, **2002**, 97, 167.

25. Grange, A.H.; Zumwalt, M.C.; Sovocool, G.W., Determination of ion and neutral loss compositions and deconvolution of product ion mass spectra using an orthogonal acceleration, time-of-flight mass spectrometer and an ion correlation program, *Rapid Commun. Mass Spectrom.*, **2006**, 20, 89–102.

26. Ferrer, I.; Fernández-Alba, A.; Zweigenbaum, J.A.; Thurman, E.M., Exact-mass library for pesticides using a molecular-feature database, *Rapid Commun. Mass Spectrom.*, **2006**, 20, 3659–3668.

27. Thurman, E.M.; Ferrer, I.; Malato, O.; Rodríguez Férnandez-Alba, A., Feasibility of LCTOFMS and elemental database searching as a spectral library for pesticides in food, *Food Addit. Contam.*, **2006**, 23, 1169–1178.

28. Ferrer, I.; Thurman, E.M., Importance of the electron mass in the calculation of exact mass by time-of-flight mass spectrometry, *Rapid Commun. Mass Spectrom.*, **2007**, 21, 2538–2539.

29. Ibáñez, M.; Pozo, O.J.; Sancho, J.V.; Hernández, F., Use of quadrupole time-of-flight mass spectrometry in environmental analysis: elucidation of transformation products of triazine herbicides in water after UV exposure, *Anal. Chem.*, **2004**, 76, 1328–1335.

30. Pozo, O.J.; Guerrero, C.; Sancho, J.V.; Ibáñez, M.; Pitarch, E.; Hogendoorn, E.; Hernández, F., Efficient approach for the reliable quantification and confirmation of antibiotics in water using on-line solid-phase extraction liquid chromatography/tandem mass spectrometry, *J. Chromatogr. A*, **2006**, 1103, 83–93.

31. Batt, A.L.; Aga, D.S., Simultaneous analysis of multiple classes of antibiotics by ion trap LC/MS/MS for assessing surface water and groundwater contamination, *Anal. Chem.*, **2005**, 77, 2940–2947.

32. Huerta-Fontela, M.; Galcerán, M.T.; Ventura, F., Ultraperformance liquid chromatography-tandem mass spectrometry analysis of stimulatory drugs of abuse in wastewater and surface waters, *Anal. Chem.*, **2007**, 79, 3821–3829.

33. Xia, Y.; Wang, P.; Bartlett, M.G.; Solomon, H.M.; Busch, K.L., An LC-MS-MS method for the comprehensive analysis of cocaine and cocaine metabolites in meconium, *Anal. Chem.*, **2000**, 72, 764–771.

CHAPTER 6

THE USE OF ACCURATE MASS, ISOTOPE RATIOS, AND MS/MS FOR THE ANALYSIS OF PPCPs IN WATER

Michael C. Zumwalt

Agilent Technologies, Inc., Englewood, Colorado

AN AGILENT 6510 Quadrupole Time of Flight Mass Spectrometer (QTOF) is used to analyze several surface water samples for the presence of pharmaceutical compounds. A simple gradient elution is carried out on an Agilent Rapid Resolution High Throughput Extend C18 column (particle size 1.8 μm). Of 54 potential compounds as many as 11 are identified in one of the samples using an algorithm known as the Molecular Feature Extractor (MFE). To make comparisons among several samples another algorithm known as Mass Profiler is applied to the data processed by the MFE. Since the MFE may generate thousands of potential compounds known as features, Mass Profiler makes statistical comparisons of the features among two different samples to determine what is unique and common. All of this work is done with the full-scan mass spectral data. When compounds of interest are determined, accurate mass full-scan MS/MS can be invoked for structural elucidation. The results of full-scan MS/MS applied to caffeine and sulfamethoxazole are included as examples and are relevant because many medications include either as ingredients.

6.1 INTRODUCTION

During the three decades prior to the year 2000 the study of chemical pollution was confined primarily to pesticides. Following a seminal article by C. Daughton [1] this focus began to shift to the emerging environmental concern for pharmaceuticals and personal care products (PPCPs). Many of these pharmaceuticals, including estrogen, have been known as endocrine disruptors, or chemicals that disrupt the physiological function of hormones in organisms [2]. In 2004 a report from the United States

Liquid Chromatography Time-of-Flight Mass Spectrometry: Principles, Tools, and Applications for Accurate Mass Analysis, Edited by Imma Ferrer and E. Michael Thurman

Geological Survey was made as a result of discovering a high preponderance of intersex (male fish exhibiting female characteristics) in Smallmouth bass of the Potomac River [3].

The USGS have found pesticides, flame retardants, and personal-care products containing known or suspected endocrine disrupting-compounds in the Potomac River and many of these compounds continue to be known as emerging contaminants as they are still being discovered and don't exist on any currently regulated target lists. As such, it is important to use adequate techniques to help identify these compounds and possible metabolites.

Using accurate mass in full-scan (mass range) mass spectrometry (MS), compound molecular formulas can be determined for purposes of identification [4]. The Agilent QTOF ensures accurate mass with the continuous introduction of two reference mass ions, which are typically at the lower and upper limits in the mass range of analysis [5]. Furthermore, the high degree of spectral resolution allows for selective identification among co-eluting compounds. Isotope ratios are an additional tool as they help identify compounds with high carbon numbers as well as those that contain elements like chlorine and sulfur. Although these tools do a lot to confirm chemical formula, it may still be left to the user to decide which of the possible molecular formulas of isobaric compounds apply.

To assist in the analytical need for structural elucidation, selective MS/MS by using the quadrupole time-of-flight mass spectrometer (QTOF) is implemented. Because the Agilent QTOF also has very accurate mass at the MS/MS level, it is easier to determine the structures of the product ions, which correspond as substructures of the precursor ion and thereby reduce the number of possible structures pertaining to the derived molecular formulas from several to one. Accurate mass at the MS/MS level is still a function of the mass analyzer flight tube control and detection of ions, which is calibrated during the MS acquisition mode. Therefore, even if only MS/MS acquisition is designated, an MS acquisition is made occasionally to maintain an excellent accurate mass range calibration.

While the principles of time-of-flight mass spectrometry, including how accurate mass and good resolution is achieved, have been presented in an earlier chapter, it is important to note the process for collisionally-induced dissociation of compound ions to produce fragment ions that allow for structural elucidation. The reason that producing product ions can be very important is that, while accurate mass may allow the investigator to reduce the possible chemical formulas to one in identifying a compound, still that single chemical formula might correspond to several different structures, or isomers. Accurate mass measurements of product ions allow the analyst to determine chemical structure.

In the Agilent 6510 QTOF instrument used for this work, the collision cell for fragmentation consists of six resistively coated rods in the presence of nitrogen buffer gas. A computer rendering is shown in Figure 6.1. The hexapole geometry is used because it provides a combination of good ion transmission over a wide mass range, essential for full-scan MS/MS, as well as good ion beam intensity for any particular m/z. The constant applied voltage gradient keeps ions moving through the collision cell. A voltage is applied on top of this to cause fragmentation. However, by always applying a minimum voltage, ions are continually being accelerated out

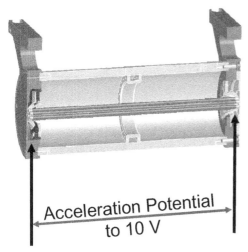

Figure 6.1.   Collision cell in QTOF to generate product ions for structural elucidation.

of the collision cell eliminating the possibility of "cross-talk." Cross-talk is the interference of product ion signal produced in an MS/MS experiment with the signal generated in a subsequent MS/MS experiment because the product ions generated earlier are still residing in the collision cell during the next fragmentation step. With a voltage gradient and a minimum inter-transition time of 5 msec in this collision cell, cross-talk is not observed.

The list of pharmaceuticals to look for in the environment is ever-increasing and many of them are metabolites with unknown structures. Identifying these compounds requires the technology of the QTOF. Furthermore, the fast-scanning capability is necessary for identifying 10s to 100s of these compounds in samples with relatively short run times. The QTOF is capable of acquiring full-scan MS data at the rate of 20 spectra/sec. The resulting large amount of data representing a possibly large number of compounds needs to be converted into useful information. The Molecular Feature Extractor (MFE), which is a standard part of the MassHunter Qualitative Analysis software, carries out the following steps:

- Persistent chemical background is removed
- Co-eluting interferences are resolved
- Isotopic clusters recognized and grouped
- 2D/3D data visualization
- Chemical identification (accurate mass, isotope matching)
- Database searching (NIST, ChemIDPlus)

In addition to applying the algorithm Molecular Feature Extractor to pull out the features from the chromatographic data, which could be compounds, another algorithm known as Mass Profiler is applied to the list of features among different samples to determine differences and commonality. Each sample is injected three

TABLE 6.1. List of Compounds with Corresponding Neutral Masses that May Be in a Given Sample

Compound	Neut. Mass	Compound	Neut. Mass	Compound	Neut. Mass
Acetaminophen	151.0633	Diphenhydramine	255.1623	Paroxetine	329.1427
Albuterol	239.1521	Duloxetine	297.1187	Ranitidine	314.1413
Aspirin	180.0423	Enalaprilat	348.1685	Sertraline	305.0738
Buproprion	239.1077	Erythromycin	733.4612	Simvastatil	418.2719
Caffeine	194.0804	Fluoxetine	309.1340	Sulfachloropyridazine	284.0135
Carbamazepine	236.0950	Fluvoxamine	318.1555	Sulfadimethoxine	310.0736
Cimetidine	252.1157	Furosemide	330.0077	Sulfamethazine	278.0838
Citalopram	324.1638	Gemifrozil	250.1570	Sulfamethizole	270.0245
Clofibric acid	214.0397	HCTZ	296.9645	Sulfamethoxazole	253.0521
Codeine	299.1521	Ketoprofen	254.0943	Thiabendazole	201.0361
Cotinine	176.0950	Miconazole	413.9860	Triclocarban	313.9780
Dehydronifedipine	344.1008	Naproxen	230.0943	Triclosan	287.9512
Diclofenac	260.0478	Norfluoxetine	295.1184	Trimethoprim	290.1379
Diltiazem	414.1613	Norsertraline	291.0582	Venlafaxine	277.2042
		1,7-dimethyl-xanthine	180.0647	Warfarin	308.1049

times or multiple samples from the same source could be used to determine what is statistically consistent in terms of the features derived for the sample by MFE. The result is called a group. Two groups representing two different sample sources can then be compared to see what features differ. Are they unique, or are they common, and if common, do they differ in abundance?

A batch of water samples are filtered and extracted using solid phase extraction, which resulted in an approximate 1000-fold increase in concentration. Samples analyzed in this work are believed to contain compounds at the 10–100 ppb level, which corresponds to the 10–100 ppt range in original water sample. The compounds that may be in these samples are included with their exact neutral masses above in Table 6.1.

6.2 EXPERIMENTAL

6.2.1 Sample Preparation

Prepared samples were provided by the United States Geological Survey National Water Quality Laboratory (USGS/NWQL) in Denver, Colorado. Pharmaceuticals are typically extracted from surface water by using disposable polypropylene syringe cartridges that contain 0.5 g of polymeric sorbent. One liter of sample is pumped through the solid-phase extraction (SPE) cartridge, which is an Oasis ® HLB cartridge (Waters Corp., Milford, MA). The analyte material is later eluted into 1 mL of methanol, resulting in a concentration increase of three orders of magnitude.

6.2.2   LC/MS Method Details

LC Conditions

Agilent 1100 series binary pump, degasser, wellplate sampler, and thermo-stated column compartment.

Column: Agilent Zorbax RRHT Extend C18, 2.1 × 50 mm, 1.8 μm (PN: 727700-902)

Column temp: 40 °C

Mobile phase: A = 0.1% formic acid in water

B = 0.1% formic acid in acetonitrile

Flow rate: 0.3 mL/min; injection vol: 5 μL

Gradient:

Time (min)	%B	
0.0	0	
10.0	67	Stop time: 15 min.
11.0	100	Post run: 10 min.

MS Conditions

Mode: positive electrospray ionization using the Agilent G3251A Dual ESI source

Nebulizer: 40 psig; Drying gas flow: 9 L/min; Drying gas temp: 350 °C

Vcap: 3500 V

Scan range: *m/z* 50–1000

Scan speed: 1 scan/sec

MS/MS Conditions

Collision energy = 30 V

Scan range: *m/z* 50–1000

Scan speed: 1 scan/sec.

6.3   RESULTS AND DISCUSSION

Of the several samples analyzed, results for Samples 4 and 10 will be considered here. To get an idea of the task at hand, an overlay of the total ion and base peak chromatograms for the first injection of Sample 4 is shown in Figure 6.2. The base peak chromatogram is generated to help the analyst identify peaks in the chromatogram corresponding to real compounds. Figure 6.3 shows the spectrum at the apex of one such peak. Note the complexity of this spectrum and the difficulty involved in not only determining which spectral peaks are of value, as they may pertain to co-eluting compounds, but then having to apply this reasoning to several peaks in the chromatogram.

Applying the algorithm of the Molecular Feature Extractor program to this data file results in the display of the processed chromatogram and the corresponding

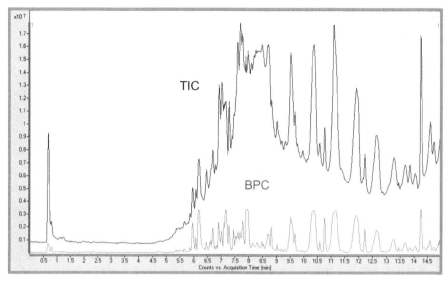

Figure 6.2. Overlay of total ion chromatogram (TIC) and base peak chromatogram (BPC) for Sample 4.

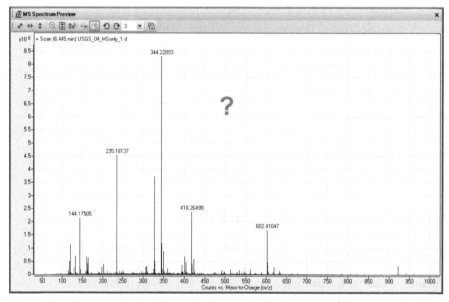

Figure 6.3. Spectrum at apex of base peak with retention time of 6.445 min.

contour plot shown in Figure 6.4. The upper left hand chromatogram is the unprocessed TIC, same as shown in Figure 6.2. On the right is the processed chromatogram after applying the steps listed in the Introduction. Random background noise has been removed. Below each of these chromatograms are shown the corresponding contour plots, which are the presentations of spectral data points in an *m/z* versus chromatographic retention time plots. The contour plot at the lower left hand corner

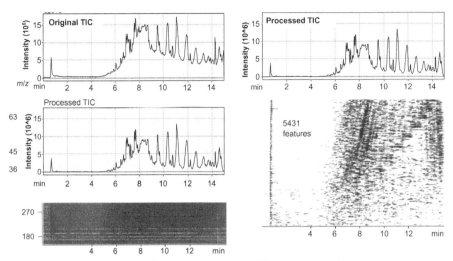

Figure 6.4.   Both unprocessed and processed data of Sample 4 using MFE.

of the display shows a very dense distribution of data points, most of which correspond to random noise.

The contour plot at the lower right hand corner is the result of processing the data so that a significant amount of molecular features are derived for closer examination. In fact, using the following settings for filtering the data, some 5431 features are derived for this sample in the first injection:

- Spectral S/N > 2
- Mass range: m/z 150–800
- $[M + Na]^+$ and $[M + NH_4]^+$ adducts considered
- Relative intensity in the spectrum > 0.1%
- Each feature must contain at least 2 ions.

If we now investigate some of the features that have been found we can begin with the peak apex spectrum examined in Figure 6.3. The retention time is 6.445 minutes and MFE has derived features at 6.448 minutes as shown in Figure 6.5. The unprocessed spectrum at top of the figure matches that of Figure 6.3. However, removing random noise and using the filtering rules above a processed spectrum containing 12 features is derived and shown at the bottom. A subset of the list of features is shown at right.

The lower spectrum (Figure 6.5) represents the processed version of the raw spectrum at the same retention time and shown above. Fewer peaks are shown in this lower spectrum because the raw spectrum has been cleaned up though processing to remove random signal and preserve only those ion masses that appear real and are also listed in the table.

If we want to filter the data to only show compounds corresponding to the list at the beginning of this article we can place the neutral masses into an inclusion list of MFE as shown in Figure 6.6. We also assume that the compounds of interest do

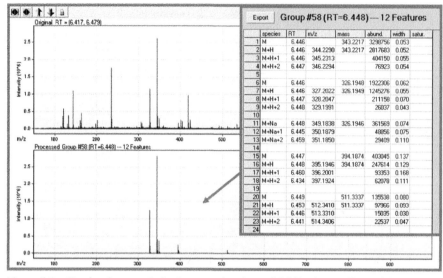

Figure 6.5. Twelve features derived at retention time of 6.448 minutes. Four of them are shown in the table.

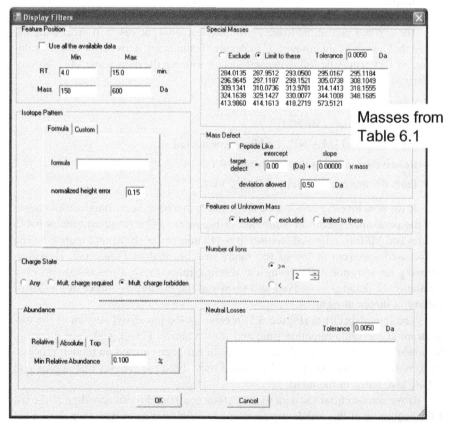

Figure 6.6. Display filter settings for finding features that match compound list of Table 6.1.

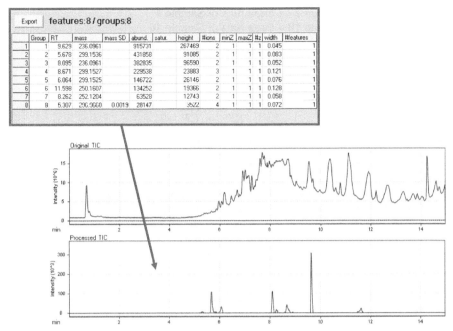

Figure 6.7.   Eight features found corresponding to the neutral masses of Table 6.1. Corresponding processed chromatogram also shown.

not elute until after 4 minutes and the mass range of interest is 150–600, which corresponds to the compounds of Table 6.1.

After applying the filtering of data with the compound list shown in Figure 6.6, eight features were found in Sample 4 as shown in Figure 6.7. The corresponding chromatogram containing these eight features is also shown.

Before looking more closely at any one of these compounds, the data of Sample 4 is now going to be compared with data from another sample, Sample 10. The comparison will be carried out using an algorithm known as Mass Profiler. In order to use Mass Profiler, at least three injections of each sample must be made to determine what is consistently there and what is random and should be disregarded. In this work each sample is injected three times. The data is first processed by MFE to generate features. Mass Profiler filters out features which are inconsistent among the three injections for each sample. The resulting data is called a Group. Therefore, in comparing Samples 4 and 10 Mass Profiler will be referring to them as Group 4 and Group 10.

In Figure 6.8, Mass Profiler shows a plot of features common to both Groups 4 and 10 and displayed as mass versus retention time. By clicking on any one of the feature points in the display one can see the common feature for both Groups along with possible molecular formulas for the derived neutral mass.

For example, in comparing features among the two sample groups a differential analysis plot can be generated as shown in Figure 6.9. In this plot, the features of Group 10 which are more or less abundant than the corresponding features in

Group 4 = three runs of Sample 4
Group 10 = three runs of Sample 10

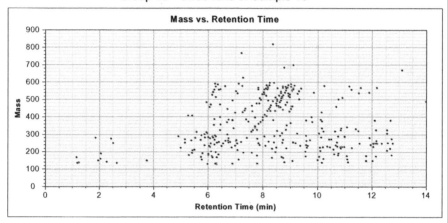

Figure 6.8.   Features present in both Groups 4 and 10 total 346.

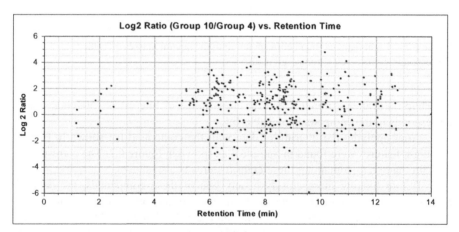

Figure 6.9.   Features common to Groups 4 and 10 but differ in magnitude.

Group 4 are represented. More specifically, at a retention time of 8.495 minutes there is a data point in Figure 6.9 that corresponds to a feature in Group 10 that is approximately 4× intensity over the corresponding feature in Group 4, which corresponds to a Log 2 Ratio of 2. By clicking on this data point in the display of Figure 6.9 one can see that this feature could be identified as diphenhydramine, with a molecular formula of $C_{17}H_{21}NO$ and accurate mass of 0.7 ppm, relative to the list of compound masses in Table 6.1. See also Figure 6.10.

It should be noted that a molecular formula doesn't reveal the molecule's structure. True confirmation would come from running a clean standard dilution of diphenhydramine and see what its retention time is. Performing accurate mass MS/

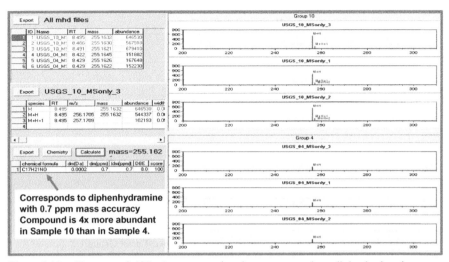

Figure 6.10.   Feature at 8.495 minutes retention time corresponds to diphenhydramine from Table 6.1.

MS would also help by determining molecular formula for both product ions and neutral losses. These molecular formulas can in turn be used to determine likely structures corresponding to pieces of the overall structure.

With Mass Profiler it is also possible to compare two samples in terms of what features are in one sample that are not in the other. In Figure 6.11 we see such a comparison. Mass Profiler has determined that there are 33 features only in Group 4 or in Group 10 and are not common to the two samples. Since the display in Mass Profiler is in color the features exclusive to Group 4 are blue and the features exclusive to Group 10 are in red. Since this chapter is published in black and white, boxes have been placed around the blue features for Group 4 for viewing convenience.

So far, all of the data has been collected in full-scan MS mode. Once features are identified as compounds needing more structural information, or it is of interest to perform some quantitation, a targeted MS/MS analysis can be performed in which the ion mass of the feature is considered as precursor ion and fragmented to form accurate mass product ions. The accurate mass of these product ions can determine their chemical formula and possible structures. Because the QTOF also has a high degree of spectral resolution in MS/MS mode, very narrow extracted ion chromatograms may be generated for each ion and then summed together for quantitation signal.

In Figure 6.12 we see the accurate mass MS/MS fragmentation of caffeine using the MS/MS settings noted in the LC/MS method details. Caffeine is of environmental interest because many medications contain it as an ingredient. Metabolites of caffeine like xanthine are also of interest, but evidence for their existence in these samples was not found. Chemical formulae for each product ion is derived based on the possible arrangements of C, H, N, and O. Knowing the structure of caffeine,

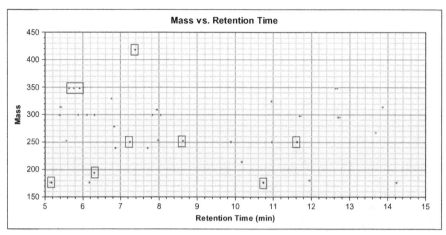

Figure 6.11. Features only in Group 4 (highlighted with boxes) or in Group 10.

Calculated chemical formula given accurate mass measurement and using elements C, H, N, and O

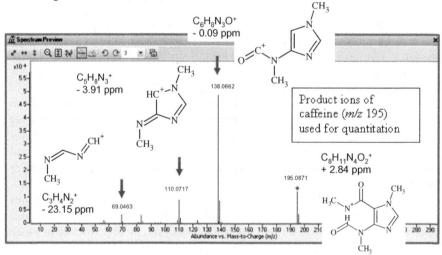

Proposed structures

Figure 6.12. Targeted MS/MS mode for caffeine producing product ions which may be used for structural elucidation as well as quantitation.

structures of the fragment ions can be proposed using their corresponding chemical formula. The fragment structures are generated using ACD/MS Fragmentor© (ACD Labs Release v. 10, Advanced Chemistry Development, Inc, Toronto, ON, Canada).

The analysis of full-scan MS/MS is repeated for the compound sulfamethoxazole, using C, H, N, O, and S. In Figure 6.13 we see the accurate mass product

Calculated chemical formula given accurate mass measurement
and using elements C, H, N, O, and S

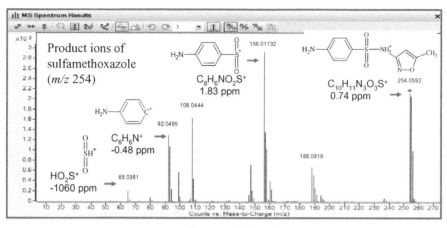

Proposed structures

Figure 6.13.   Targeted MS/MS mode for sulfamethoxazole producing product ions which
may be used for structural elucidation as well as quantitation.

ions and the precursor ion all having accuracies of 2 ppm or less, except for the
product ion at m/z 65.0381, which has a very large mass error if it corresponds to
the fragment ion shown. A contributing factor to this mass error is the fact that the
reference ions used for fine calibration of the mass axis are at m/z values of 121.0509
and 922.0098. These ions are sprayed into the ion source along with the LC effluent
in order to maintain excellent mass accuracy at least among those ions within the
mass range between these reference ions. Since the m/z 65.0381 ion is outside this
range, its measured accuracy will not be as good.

6.4   CONCLUSIONS

The QTOF is an excellent instrument for identifying compounds using accurate mass
in full-scan MS and MS/MS. Accurate mass leads to chemical formula, which can
also give structural information when forming product ions in MS/MS. As a lot of
data is acquired by this type of instrument to look at samples that may contain large
amounts of known and unknown compounds, it is important to have algorithms like
Molecular Feature Extractor that can filter usable features out of the chemical background. These features are generated from spectra as a result of removing random
background signal and finding clusters of isotopes that make sense.

While this analysis is useful for one sample it may also be important to make
comparisons among multiple samples as well. Another algorithm known as Mass
Profiler makes such comparisons. More specifically, comparisons such as what is
common to two samples and how do they differ in amount? Or, what features are

in one sample that aren't in the other. Once the feature is considered for more investigation, targeted MS/MS may be carried out on that feature to get structural information based on the generation of product ions.

ACKNOWLEDGMENTS

The author gratefully acknowledges the assistance of Stephen Werner and Ed Furlong of the National Water Quality Lab–United States Geological Survey (Lakewood, CO) for providing the samples analyzed in this work, and his Agilent colleagues, David Weil and Chin-Kai Meng for their expert advice.

REFERENCES

1. Daughton, C.G.; Ternes, T.A., Pharmaceuticals and personal care products in the environment: agents of subtle change? *Environ. Health Perspect.*, **1999**, 107, Suppl. 6, 907–938.
2. Krimsky, S., An epistemological inquiry into the endocrine disruptor, thesis, *Ann. N.Y. Acad. Sci.*, **2005**, 948, 130–142.
3. Chambers, D.B.; Leiker, T.J., A Reconnaissance for Emerging Contaminants in the South Branch Potomac River, Cacapon River, and Williams River Basins, West Virginia, April–October 2004, Open File Report 2006-1393, United States Geological Survey, http://pubs.usgs.gov/of/2006/1393/.
4. McIntyre, D., Using the Agilent LC/MSD TOF to Identify Unknown Compounds, Agilent Technologies Application Note., 2004, publication number 5989-0626EN.
5. Fjeldsted, J.C., Time-of-Flight Mass Spectrometry, Agilent Technologies Technical Overview, 2003, publication number 5989-0626EN.

PART III

APPLICATIONS OF LC/TOF-MS FOR THE IDENTIFICATION OF SMALL MOLECULES

APPLICATIONS OF LC/TOF-MS IN THE ENVIRONMENTAL FIELD: WHEN DID IT START AND WHERE IS IT GOING?

Imma Ferrer and E. Michael Thurman

Center for Environmental Mass Spectrometry, Department of Civil, Environmental and Architectural Engineering, University of Colorado, Boulder, Colorado

THE ADVENT of time-of-flight techniques applied to environmental analyses has just begun in the last few years. Applications range from routine analytical methods that analyze a few target compounds to more extensive methods that include a variety of analytes, including also non-target and unknown identification. Due to the high complexity of some environmental samples (i.e., wastewater, sludge samples, food samples), high-resolution techniques with additional structural information on fragment ions are needed. These techniques provide a high degree confidence for identification of target analytes and aid to the structural elucidation of degradation products and unknown compounds, which are usually present in environmental samples. The possibility of creating universal accurate mass databases with time-of-flight analyses for sets of compounds has broadened the range of applications as well, going from target to non-target identification. This chapter gives an overview of the starting point for time-of-flight techniques and the applications that have recently generated in the environmental field. Furthermore, it states the acceptance of this relatively "new technique" among the environmental scientific community by reviewing some of the key achievements and challenges on this topic.

7.1 EMERGING CONTAMINANTS ISSUE

In the last few years, the analysis of emerging contaminants, many of which were unknown until recently, has been the main focus of environmental scientists [1]. Emerging contaminant issues have been highlighted by several scientific meetings

Liquid Chromatography Time-of-Flight Mass Spectrometry: Principles, Tools, and Applications for Accurate Mass Analysis, Edited by Imma Ferrer and E. Michael Thurman
Copyright © 2009 John Wiley & Sons, Inc.

and a series of papers and reviews that deal with these compounds in the environment [2, 3]. In the last ten years an important finding has been the presence of widespread pharmaceuticals in surface water in Europe [4–6] and in the United States [7]. The results of one of the reconnaissances by the U.S. Geological Survey showed that 80% of all surface water had detectable concentrations of pharmaceutical compounds. Approximately 82 compounds were detected including steroids, antibiotics, analgesics, heart medications, and other compounds. Typically the concentrations were in the sub-microgram-per-liter range. The majority of the pharmaceuticals identified were detected using liquid chromatography/mass spectrometry [7]. Another topic of interest has been the detection of pesticides in food samples. In this sense, various recent papers have been published showing the importance of conducting this type of analyses in food matrices [8–10]. A trend has been noticed in the last few years as the application of LC-MS techniques to environmental problems have increased exponentially.

7.2   LC-MS TECHNIQUES AND THE ADVENT OF LC/TOF-MS

Identification of pharmaceutical compounds [2] and new pesticide degradates [11] in the environment are two excellent examples that have followed the advent of LC/MS instrument technology. In many cases, unequivocal identification by GC/MS has not been possible, mainly because the compounds are not volatile in the inlet of the GC. LC/MS, on the other hand, will ionize many compounds in either electrospray or atmospheric pressure chemical ionization. In this case, identification typically involves two steps. Firstly, there is the production of a mass spectrum from an authentic standard to create a user library. Secondly, there is identification by matching chromatographic retention time and mass spectrum for molecular ion and fragment ions. This procedure was considered adequate for final identification until recently, when this level of identification has been challenged for complex samples containing many compounds. Because LC/MS relies on fragmentation using collision induced dissociation (CID) in the source of the mass spectrometer, co-eluting peaks in the mass spectrum will interfere with each other resulting in multiple CID spectra that are superimposed on one another. LC/MS-MS removes this problem and results in a much higher degree of certainty in identification of unknowns. However, with LC/MS-MS, there still exists the problem of identification of an unknown peak from only the mass spectrum. Because there are no libraries of non-targets or unknowns available at this time, except of course for user created libraries, it is difficult to identify a compound from only a mass spectrum. The use of accurate mass, both with time-of-flight coupled to mass spectrometry (TOF-MS) and with sector instruments, have been the solution to this problems of unknown identification. Examples are given in this book where the authors have addressed this problem with a series of analytical approaches that allow the identification of compounds that are true unknowns.

It is important to note the evolution of time-of-flight techniques applied to analytical problems over the years. Figure 7.1 shows the number of papers published

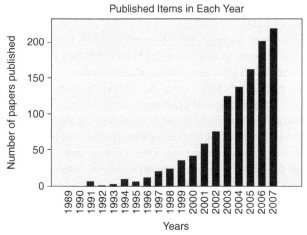

Figure 7.1.   Trend in the total number of applications for LC/TOF-MS in the environmental field.

when doing a search in the "web of science" page that Thomson Scientific publishes on the web. Our search was mainly for a combination of TOF and LC techniques, excluding all the hits that contained MALDI since this book is mainly focused on LC/TOF-MS techniques. The results are depicted in Figure 7.1 and show the increasing trend of TOF techniques applied to environmental and medical problems. The use of these techniques started a little over 10 years ago and the applications have been exponentially increasing in the last five years. The total number of hits we came up with (at the time this book was submitted) was for a total of 1,175 papers using LC time-of-flight techniques.

7.3   APPLICATIONS OF LC/TOF-MS IN THE ENVIRONMENTAL FIELD

Recently, LC/TOF-MS has been used for the unequivocal confirmation of contaminants (including pharmaceuticals, pesticides and surfactants) in a variety of samples, such as water, sediments and food matrices [12, 13] by accurate mass measurement of protonated molecules. In this sense, several authors have reported accurate mass confirmation of pharmaceuticals in surface and wastewater samples [14–18] as well as sediment and sludge [19] using time-of-flight techniques. Detection of drugs in urine has also been one of the topics that have been recently covered by LC/TOF-MS techniques [20–23]. The same trend has been observed for the identification of pesticides and their degradation products in water samples [24–27]. In many of these studies time-of-flight techniques were successfully used for the unequivocal identification of degradation products of known pesticides, as well as unknown compounds [24, 27]. The application of time-of-flight to food commodities has also experienced an increase in the last few years [8, 9, 27–32]. It is worthy to mention also several applications of time-of-flight mass analysis for the identification and confirmation of metabolites or degradation products of pesticides and pharmaceuticals in environmental samples [33–42].

Accurate mass measurements of fragment ions, become particularly important in the structure elucidation of unknowns. In this sense, the Q-TOF MS/MS is unique among TOF instruments in its ability to give accurate mass measurements (1–2 millimass units) of the fragment ions that are ejected from the collision chamber, which give a high assurance of correct identification of unknowns, as well as an empirical formula of fragment ions. Sometimes, a single stage time-of-flight mass analyzer (TOF/MS) generates this information by imparting enough energy into the $[M + H]^+$ ions in the source region to cause fragmentation [43]. Time-of-flight mass analysis generates increased resolving power of signals on the m/z axis in comparison to quadrupole mass spectrometers. Furthermore, this enhanced resolving power benefits analyses involving complex environmental matrices, by separating isobaric interferences from the contaminant signals of interest. The improved resolution also facilitates the measurement of accurate masses within 5 ppm, which are accepted for the verification of elemental compositions. Elemental compositions of contaminants and their fragment ions clearly constitute higher order identifications than those afforded by nominal mass measurements.

Another important tool that has made TOF one of the key methodologies for identification of compounds is the existence of accurate mass databases, as seen in Chapter 4. An individual scientist can apply these universal databases to each particular problem and then often get an identification on the analyte of interest [44–46].

This section gives several examples of emerging-contaminant analysis that exemplify the unique features of time-of-flight for the identification of target and unknown compounds. Liquid chromatography/mass spectrometry employing accurate mass measurement has been proved as a successful technique for quantitative analysis of target compounds and rapid qualitative analysis of "unknown" environmental mixtures.

REFERENCES

1. Richardson, S.D., Water analysis: emerging contaminants and current issues, *Anal. Chem.*, **2007**, 79, 4295–4323.
2. Daughton, C.G.; Jones-Lepp, T.L. (Editors), Pharmaceuticals and Personal Care Products in the Environment, American Chemical Society Symposium Series 791, Oxford University Press, New York, **2001**.
3. Daughton, C.G.; Ternes, T.A., Pharmaceuticals and personal care products in the environment: agents of subtle change? *Environ. Health Perspect.*, **1999**, 107, 907–938.
4. Daughton, C.G.; Ternes, T.A., Occurrence of drugs in German sewage treatment plants and rivers, *Water Research*, **1998**, 32, 3245–3260.
5. Ternes, T.; Anderson, H.; Gilberg, D.; Bonerz, M., Determination of estrogens in sludge and sediments by liquid extraction and GC/MS/MS, *Anal. Chem.*, **2002**, 74, 3498–3504.
6. Hirsch R.; Ternes, T.; Haberer, K.; Kratz, K.L., Occurrence of antibiotics in the aquatic environment, *Sci. Total Environ.*, **1999**, 225, 109–118.
7. Kolpin, D.W.; Furlong, E.T.; Meyer, M.T.; Thurman, E.M.; Zaugg, S.D.; Barber, L.B.; Buxton, H.T., Pharmaceuticals, hormones, and other organic wastewater contaminants in US streams, 1999–2000: A national reconnaissance, *Environ. Sci. Technol.*, **2002**, 36, 1202–1211.

8. Ferrer, I.; García-Reyes, J.F.; Mezcua, M.; Thurman, E.M.; Fernández-Alba, A.R., Multi-residue pesticide analysis in fruit and vegetables by liquid chromatography-time-of-flight-mass spectrometry, *J. Chromatogr. A*, **2005**, 1082, 81–90.

9. Thurman, E.M.; Ferrer, I.; Zweigenbaum, J.A., High resolution and accurate mass analysis of xenobiotics in food, *Anal. Chem.*, **2006**, 78, 6702–6708.

10. Pico, Y.; Blasco, C.; Font, G., *Mass Spectrom. Rev.*, Environmental and food applications of LC-tandem mass spectrometry in pesticide-residue analysis: an overview, **2004**, 23, 45–85.

11. Meyer, M.T.; Thurman, E.M. (Editors), Herbicide Metabolites in Surface and Groundwater; ACS Symposium Series 630; American Chemical Society: Washington, **1996**.

12. Ferrer, I.; Thurman, E.M. (Editors), Liquid Chromatography Mass Spectrometry/Mass Spectrometry, MS/MS and Time-of-Flight MS: Analysis of Emerging Contaminants, American Chemical Society Symposium Series 850, Oxford University Press, New York, **2003**.

13. Ferrer, I.; Thurman, E.M., Liquid chromatography/time-of-flight/mass spectrometry (LC/TOF/MS) for the analysis of emerging contaminants, *Trends Anal. Chem.*, **2003**, 22, 750–756.

14. Farre, M.; Gros, M.; Hernandez, B.; Petrovic, M.; Hancock, P.; Barceló, D., Analysis of biologically active compounds in water by ultra-performance liquid chromatography quadrupole time-of-flight mass spectrometry, *Rapid Commun. Mass Spectrom.*, **2008**, 22, 41–51.

15. Radjenovic, J.; Petrovic, M.; Barceló, D., Title: Advanced mass, spectrometric methods applied to the study of fate and removal of pharmaceuticals in wastewater treatment, *TRAC Trends Anal. Chem.*, **2007**, 26, 1132–1144.

16. Ibáñez, M.; Sancho, J.V.; McMillan, D.; Rao, R.; Hernández, F., Rapid non-target screening of organic pollutants in water samples by ultra performance liquid chromatography-time of flight mass spectrometry, *TRAC Trends Anal. Chem.*, **2008**, 27, 481–489.

17. Gómez, M.J.; Malato, O.; Ferrer, I.; Agüera, A.; Fernández-Alba, A.R., Solid-phase extraction followed by liquid chromatography–time-of-flight–mass spectrometry to evaluate pharmaceuticals in effluents. A pilot monitoring study, *J. Environ. Monitor.*, **2007**, 9, 718–729.

18. Stolker, A.A.M.; Niesing, W.; Hogendoorn, E.A.; Versteegh, J.F.M.; Fuchs, R.; Brinkman, U.A.Th., Liquid chromatography with triple-quadrupole or quadrupole-time of flight mass spectrometry for screening and confirmation of residues of pharmaceuticals in water, *Anal. Bioanal. Chem.*, **2004**, 378, 955–963.

19. Ferrer, I.; Heine, C.E.; Thurman, E.M., Combination of LC/TOF/MS and LC/Ion trap/MS/MS for the identification of diphenhydramine (Benadryl) in sediment samples, *Anal. Chem.*, **2004**, 76, 1437–1444.

20. Kaufmann, A.; Butcher, P.; Maden, K.; Widmer, M., Ultra-performance liquid chromatography coupled to time of flight mass spectrometry (UPLC-TOF): a novel tool for multiresidue screening of veterinary drugs in urine. *Anal. Chim. Acta.*, **2007**, 586, 13–21.

21. Ojanperä, I.; Pelander, A.; Laks, S.; Gergov, M.; Vuori, E.; Witt, M., Application of accurate mass measurement to urine drug screening, *J. Anal. Toxicol.*, **2005**, 29, 34–40.

22. Ojanperä, S.; Pelander, A.; Pelzing, M.; Krebs, I.; Vuori, E.; Ojanperä, I., Isotopic pattern and accurate mass determination in urine drug screening by liquid chromatography/time-of-flight mass spectrometry, *Rapid Commun. Mass Spectrom.*, **2006**, 20, 1161–1167.

23. Pelander, A.; Ojanperä, I.; Laks, S.; Rasanen, I.; Vuori, E., Toxicological screening with formula-based metabolite identification by liquid chromatography/time-of-flight mass spectrometry, *Anal. Chem.*, **2003**, 75, 5710–5718.

24. Ibáñez, M.; Sancho, J.V.; Pozo, O.J.; Niessen, W.M.A.; Hernández, F., Use of quadrupole time-of-flight mass spectrometry in the elucidation of unknown compounds present in environmental water, *Rapid Commun. Mass Spectrom.*, **2005**, 19, 169–178.

25. Sancho, J.V.; Pozo, O.J.; Ibáñez, M.; Hernández, F., Potential of liquid chromatography/time-of-flight mass spectrometry for the determination of pesticides and transformation products in water, *Anal. Bioanal. Chem.*, **2006**, 386, 987–997.

26. Ibáñez, M.; Pozo, O.J.; Sancho, J.V.; Hernández, F., Use of quadrupole time-of-flight mass spectrometry in environmental analysis: elucidation of transformation products of triazine herbicides in water after UV exposure, *Anal. Chem.*, **2004**, 76, 1328–1335.

27. Thurman, E.M.; Ferrer, I.; Fernández-Alba, A.R., Matching unknown empirical formulas to chemical structure using LC/MS TOF accurate mass and database searching: examples of unknown pesticides on tomato skins, *J. Chromatogr. A*, **2005**, 1067, 127–134.

28. Ferrer, I.; Thurman, E.M., Multi-residue method for the analysis of 101 pesticides and their degradates in food and water samples by liquid chromatography/time-of-flight mass spectrometry, *J. Chromatogr. A*, **2007**, 1175, 24–37.

29. Grimalt, S.; Pozo, O.J.; Sancho, J.V.; Hernandez, F., Use of liquid chromatography coupled to quadrupole time-of-flight mass spectrometry to investigate pesticide residues in fruits, *Anal. Chem.*, **2007**, 79, 2833–2843.

30. Wang, J.; Leung, D., Analyses of macrolide antibiotic residues in eggs, raw milk, and honey using both ultra-performance liquid chromatography/quadrupole time-of-flight mass spectrometry and high-performance liquid chromatography/tandem mass spectrometry, *Rapid Commun. Mass Spectrom.*, **2007**, 21, 3213–3222.

31. Calbiani, F.; Careri, M.; Elviri, L.; Mangia, A.; Zagnoni, I., Accurate mass measurements for the confirmation of Sudan azo-dyes in hot chilli products by capillary liquid chromatography-electrospray tandem quadrupole orthogonal-acceleration time of flight mass spectrometry, *J. Chromatogr. A*, **2004**, 1058, 127–135.

32. Soler, C.; Hamilton, B.; Furey, A.; James, K.J.; Mañes, J.; Picó, Y., Liquid chromatography quadrupole time-of-flight mass spectrometry analysis of carbosulfan, carbofuran, 3-hydroxycarbofuran, and other metabolites in food, *Anal. Chem.*, **2007**, 79, 1492–1501.

33. Thurman, E.M.; Ferrer, I.; Zweigenbaum, J.A.; García-Reyes, J.F.; Woodman, M.; Fernández-Alba, A.R., Discovering metabolites of post-harvest fungicides in citrus with liquid chromatography/time-of-flight mass spectrometry and ion trap tandem mass spectrometry, *J. Chromatogr. A*, **2005**, 1082, 71–80.

34. Picó, Y.; Farré, M.; Soler, C.; Barceló, D., Identification of unknown pesticides in fruits using ultra-performance liquid chromatography-quadrupole time-of-flight mass spectrometry imazalil as a case study of quantification, *J. Chromatogr. A*, **2007**, 1176, 123–134.

35. Ibañez, M.; Sancho, J.V.; Pozo, O.J.; Hernandez, F., Use of liquid chromatography quadrupole time-of-flight mass spectrometry in the elucidation of transformation products and metabolites of pesticides. Diazinon as a case study, *Anal. Bioanal. Chem.*, **2006**, 384, 448–457.

36. Picó, Y.; Farré, M.; Soler, C.; Barceló, D., Confirmation of fenthion metabolites in oranges by IT-MS and QqTOF-MS, *Anal. Chem.*, **2007**, 79, 9350–9363.

37. Agüera, A.; Pérez Estrada, L.A.; Ferrer, I.; Thurman, E.M.; Malato, S.; Fernández-Alba, A.R., Application of time-of-flight mass spectrometry to the analysis of phototransformation products of diclofenac in water under natural sunlight, *J. Mass Spectrom.*, **2005**, 40, 908–915.

38. Ferrer, I.; Mezcua, M.; Gomez, M.J.; Thurman, E.M.; Aguera, A.; Hernando, M.D.; Fernandez-Alba, A.R., Liquid chromatography/time-of-flight mass spectrometric analyses for the elucidation of the photodegradation products of triclosan in wastewater samples, *Rapid Commun. Mass Spectrom.*, **2004**, 18, 443–450.

39. Lambropoulou, D.A.; Hernando, M.D.; Konstantinou, I.K.; Thurman, E.M.; Ferrer, I.; Albanis, T.A.; Fernandez-Alba, A.R., Identification of photocatalytic degradation products of bezafibrate in TiO(2) aqueous suspensions by liquid and gas chromatography, *J. Chromatogr. A*, **2008**, 1183, 38–48.

40. Mezcua, M.; Ferrer, I.; Hernando, M.D.; Fernández-Alba, A.R., Photolysis and photocatalysis of Bisphenol A: identification of degradation products by liquid chromatography with electrospray ionization/time-of-flight/mass spectrometry (LC/ESI/ToF/MS), *Food Additives and Contaminants*, **2006**, 23, 1242–1251.

41. Perez, S.; Eichhorn, P.; Barceló, D.; Aga, D.S., Structural characterization of photodegradation products of enalapril and its metabolite enalaprilat obtained under simulated environmental conditions by hybrid quadrupole-linear ion trap-MS and quadrupole-time-of-flight-MS, *Anal. Chem.*, **2007**, 79, 8293–8300.

42. Perez, S.; Farkas, M.; Barceló, D.; Aga, D.S., Characterization of glutathione conjugates of chloroacetanilide pesticides using ultra-performance liquid chromatography/quadrupole time-of-flight mass spectrometry and liquid chromatography/ion trap mass spectrometry, *Rapid Commun. Mass Spectrom.*, **2007**, 21, 4017–4022.

43. Thurman, E.M.; Ferrer, I.; Pozo, O.J.; Sancho, J.V.; Hernández, F., The even-electron rule in electrospray mass spectra of pesticides, *Rapid Commun. Mass Spectrom.*, **2007**, 21, 3855–3868.

44. Ferrer, I.; Fernández-Alba, A.R.; Zweigenbaum, J.A.; Thurman, E.M., Exact-mass library for pesticides using a molecular-feature database, *Rapid Commun. Mass Spectrom.*, **2006**, 20, 3659–3668.
45. Thurman, E.M.; Ferrer, I.; Malato, O.; Fernández-Alba, A.R., Feasibility of LC/TOFMS and elemental database searching as a spectral library for pesticides in food, *Food Additives and Contaminants*, **2006**, 23, 1169–1178.
46. Polettini, A.; Gottardo, R.; Pascali, J.P.; Tagliaro, F., Implementation and performance evaluation of a database of chemical formulas for the screening of pharmaco/toxicologically relevant compounds in biological samples using electrospray ionization-time-of-flight mass spectrometry, *Anal. Chem.*, **2008**, 80, 3050–3057.

CHAPTER 8

LIQUID CHROMATOGRAPHY AND AMBIENT IONIZATION TIME-OF-FLIGHT MASS SPECTROMETRY FOR THE ANALYSIS OF GENUINE AND COUNTERFEIT PHARMACEUTICALS

Facundo M. Fernández*, Christina Y. Hampton,
Leonard Nyadong, Arti Navare, and Mark Kwasnik

School of Chemistry and Biochemistry, Georgia Institute of Technology,
Atlanta, Georgia

AN **OVERVIEW** of the use of liquid chromatography (LC) and several recently developed "ambient" ionization techniques is given with an emphasis on the coupling of these methods to high resolution mass spectrometers to obtain accurate mass data. The advantages of using accurate mass spectrometry (MS) are highlighted by two case studies investigating the widespread problem of drug counterfeiting using either LC-MS or direct analysis in real time (DART), a high-throughput ambient ionization method. In the first study, LC-MS was used to characterize genuine and counterfeit antimalarial pharmaceuticals labeled as containing artesunate. Of 30 tablets analyzed, 68% were found to be counterfeit and a small subset (35%) contained wrong active ingredients which were identified by using accurate mass measurements. In the second study, DART-MS was used to identify pharmaceuticals in low quality combination packets sold along the Thai-Myanmar border. A set of 58 packets containing 182 tablets or capsules were screened and the active ingredients in 91% of the samples were successfully identified by combining accurate mass information with isotope ratio matching. These case studies demonstrate the

Liquid Chromatography Time-of-Flight Mass Spectrometry: Principles, Tools, and Applications
for Accurate Mass Analysis, Edited by Imma Ferrer and E. Michael Thurman
Copyright © 2009 John Wiley & Sons, Inc.

versatility and analytical power of high-resolution mass spectrometers, especially when coupled to the latest generation of high-throughput ionization techniques.

8.1 INTRODUCTION

The advent of the "new time-of-flight" mass spectrometry (MS) [1], catalyzed by technological advances in reflectron technology, collisional cooling, orthogonal ion injection, hybrid time-of-flight instrumentation, and analog-to-digital data acquisition has resulted in a new generation of MS detectors for liquid chromatography-mass spectrometry (LC-MS) with improved mass resolution (6000–20,000), mass accuracy (10–2 ppm), dynamic range (4–5 orders of magnitude), duty cycle (15–100%) and sensitivity (fmol). This revolution in MS instrumentation has put an exquisitely powerful analytical tool at the chromatographer's fingertips, enabling the simultaneous non-targeted detection, identification, and quantitation of hundreds of analytes.

In recent years, emerging direct ionization technologies have significantly expanded the toolbox of analytical methodologies available for drug discovery, metabolomics, forensics, and quality control applications. This has allowed to first screen samples without requiring any preparation or chromatographic separation, which can be followed by a more comprehensive LC-MS examination, if necessary. In this two-tiered approach, samples are first screened in a high-throughput fashion to establish an estimate of their complexity, analyte concentration range, the presence of unknowns, and authenticity, following which, samples of interest, as determined by the high-throughput screening are then subjected to more broad, but time-consuming, LC-MS analysis for in-depth chemical characterization. This new family of ionization methods which require no sample preparation, known collectively as "ambient," "open-air," or "direct" ionization techniques, is revolutionizing mass spectrometry by enabling the rapid screening of chemical systems of medium complexity [2] with sample throughputs of up to 45 samples min^{-1} [3].

Ambient ionization methods are characterized by operating at atmospheric pressure, and by being able to probe the surface of samples of any size, shape, and texture/morphology. Ions are created outside of the instrument, in the open air. Whereas in conventional mass spectrometric experiments the sample is generally dissolved prior to analysis, in ambient MS the sample is interrogated in its native state, accelerating the analytical pipeline, preserving spatial chemical information, avoiding dilution, and maximizing sensitivity. The breadth of the new field of ambient MS is reflected by the explosive appearance of a multitude of new ionization approaches, spearheaded by desorption electrospray ionization (DESI) [4], and followed by direct analysis in real time (DART) [5], desorption sonic spray ionization (DeSSI) [6], neutral desorption sampling extractive electrospray ionization (ND-EESI) [7], desorption atmospheric pressure chemical ionization (DAPCI) [8], plasma-assisted desorption/ionization (PADI) [9], dielectric barrier discharge ionization (DBDI) [10], atmospheric-pressure solids analysis probe (ASAP) [11], electrospray-assisted laser desorption ionization (ELDI) [12], matrix-assisted laser desorption electrospray ionization (MALDESI) [13], and laser ablation-electrospray ionization (LAESI) [14].

Here, we provide a brief overview on the principles and applications of existing ambient ionization technologies, emphasizing those which have been coupled to high resolution mass spectrometers for accurate mass experiments. We then summarize our own experience following two case studies. First, we summarize the findings of forensic epidemiological studies of large-scale cases of drug counterfeiting, such as those found in Southeast Asia and Africa, utilizing accurate mass LC-MS. Secondly, we describe the results of a drug quality survey carried out using DART-MS in regions of developing countries where malaria is endemic.

8.2 SPRAY OR JET-BASED AMBIENT IONIZATION TECHNIQUES: DESI, DeSSI

Desorption electrospray ionization (DESI), first introduced by the Cooks group [4] in 2004, makes use of a pneumatically assisted high-velocity electrospray jet which is directed onto the sample surface to desorb the condensed-phase analyte(s). Following desorption, charged analyte-containing secondary droplets are sampled by a mass spectrometer. DESI enables the analysis of materials over a large mass range, and provides the capability to perform ion-molecule reactions at the interface between the charged microdroplets and the condensed-phase analytes, resulting in improved detection, selectivity and sensitivity [15].

Several ionization mechanisms were initially proposed for DESI [4, 16]: a) "splashing" on the sample surface followed by droplet pick-up, b) heterogeneous reactions between gas phase ions and surface molecules ("chemical sputtering") [17, 18], and c) desorption of neutral species from the surface due to momentum transfer, followed by gas phase ionization through ion/molecule reactions. More recent experimental studies with Doppler particle sizing analysis [19] have shown the kinetic energy per impacting water molecule to be less than $0.6 \, \text{meV}$, and thus, sputtering through momentum transfer during collisions or ionization by other electronic processes is now considered unlikely. During the course of these experiments, it was also observed that some droplets appeared to roll along the surface, increasing contact time and presumably the amount of material that is taken up into droplets during conditions typical of the DESI experiment. It was then proposed that analyte pick-up by impacting droplets takes place during the brief contact time when droplets collide with the sample surface, after which, ionization occurs by ion evaporation or charge residue mechanisms [20]. Recently, computational fluid dynamics simulations have confirmed this "droplet pick-up mechanism" showing that the DESI process follows three consecutive stages: a) the formation of a thin liquid film on the sample surface, b) extraction of the solid-phase analytes into this thin film, and c) collision of primary droplets with this thin film, producing secondary droplets which take up part of the analyte-containing liquid [21].

In a related technique, namely, desorption sonic spray ionization (DeSSI), a high velocity spray is used to probe the sample, but with the difference that no voltage is applied to the sprayed solution, thus providing a voltage-free ionization method. This approach has been shown to be useful for the direct analysis of pharmaceuticals [6]. It is believed that in DeSSI, droplet charging occurs due to statistical fluctuations of the charge spatial distributions within the sprayed droplets following the sonic spray mechanism [22–24]. Upon collision with the sample surface, these droplets probably produce analyte desorption and ionization following the DESI mechanisms described above.

One particularly interesting and somewhat unexplored DESI aspect is its "reactive" mode. In reactive mode, specific chemical reactions are achieved by the introduction of various reagents into the DESI spray solution [15, 25, 26]. Cooks and coworkers were the first to demonstrate the potential of reactive DESI for the trace detection of explosives from a wide variety of surfaces in very short times (<5 s) by selective adduction with chloride, trifluoroacetate, and methoxide anions introduced in the DESI spray [26]. In reactive DESI, the mass spectrometric detection of specific products of these *in-situ* chemical reactions is used as an added layer of selectivity to the already specific nature of the MS detector or as an aid in analyte ionization. In some other cases, the reactions conducted during the reactive DESI process produce chemical species which more efficiently survive transport through the mass spectrometer's ion optics resulting in an improvement in sensitivity [15].

Due to the relative familiarity of the instrumentation involved, and the potential of this technique, DESI is being rapidly adopted by a growing community of researchers. Table 8.1 describes a variety of applications that have been successfully tackled by DESI-MS, which include several that make use of high-resolution mass spectrometers [27–32].

8.3 ELECTRIC DISCHARGE-BASED AMBIENT IONIZATION TECHNIQUES: DART, ASAP, DAPCI, DBDI, AND PADI

Direct Analysis in Real Time (DART) was first introduced in 2005 [5] by Cody et al. and has quickly gained popularity among MS practitioners. At the core of the DART ion source is a point-to-plane corona discharge supported by a He stream. A series of complex processes within this discharge (electron-impact, ion-electron recombination), produce metastable He atoms (He*, $^3S_1$, 19.8 eV), which are carried downstream by the He gas flow. Nitrogen can be used instead of He, in which case, the excited species correspond to energetic N_2 vibronic states. Ions and electrons, also produced in the corona discharge, are filtered out by a grid electrode. The He*-containing gas stream is heated to temperatures varying from 200 to 450 °C prior to exiting the ion source, aiding in analyte desorption. A second grid electrode at the exit of the ion source reduces the extent of ion-ion recombination within the open air ionization region [5].

Cody et al. have proposed that one of several possible DART ionization mechanisms follows equations (1–3), where, upon exiting the ion source, He metastables induce Penning ionization [33] of atmospheric water, generating protonated water clusters [5] (eq. 1). The analyte(s) (*AB*), vaporized from the sample by the

TABLE 8.1.   DESI-MS Applications

Year	Authors	Ionization Technique Used	Application	Comments	Reference
2005	Cooks et al.	DESI	Pharmaceuticals (tablets, ointments, and liquids)	Focus on high-throughput analysis and effect of experimental variables	[3]
2006	Cooks et al.	DESI	Tissue imaging	Brain tissue sections with mostly lipids detected	[66]
2007	Cooks et al.	DESI	Novel DESI interface	DESI performance independent of DESI configuration	[67]
2007	Cooks et al.	DESI	Non-proximate detection	Small and large molecules are detected at distances up to 3 m	[68]
2007	Cooks et al.	DESI	Metabolomics	Urine metabolites	[69]
2006	Cooks et al.	DESI	DESI on portable mass spectrometer	Various small molecules including pharmaceuticals, explosives, plant tissues, and chemical warfare simulants	[70]
2005	Cooks et al.	DESI	Explosives	Different surfaces and ionization modes tested	[8]
2006	Cooks et al.	DESI	Industrial polymers	Electrosonic spray ionization and DESI compared	[71]
2005	Cooks et al.	DESI	Tissue imaging	Phospholipids primarily detected	[72]
2006	Cooks et al.	DESI	Metabolomics	Nuclear magnetic resonance and DESI combined	[73]
2006	Cooks et al.	DESI	High resolution measurements	DESI coupled to Orbitrap-MS	[27]
2005	Cooks et al.	DESI	Various	Review article	[16]
2007	Cooks et al.	DESI	Untreated bacteria	Microbial identification by fingerprinting and multivariate analysis	[74]
2006	Cooks et al.	DESI	Explosives and chemical warfare agents	Non-proximate detection	[75]
2007	Cooks et al.	DESI	Forensic imaging	Different inks were examined directly from ordinary paper	[76]
2007	Kostiainen, Cooks et al.	DESI	Drugs of abuse and their metabolites	Comparison of DESI and GC-MS results from urine samples	[77]
2007	Cooks et al.	DESI	Detection of explosives on skin	TNT, RDX, HMX, PETN detected	[78]
2006	Cooks et al.	Reactive DESI	Cis-diol functionality	Selective complexation reaction to form a cyclic boronate	[79]
2005	Cooks et al.	Reactive DESI	Explosives on surfaces	Chloride and trifluoroacetate adducts of RDX and HMX and Meisenheimer complex of TNT detected	[26]
2007	Cooks et al.	Reactive DESI	Phosphonate esters	Boric acid added to spray solvent for increased selectivity	[80]
2006	Cooks et al.	Reactive DESI	Copper(II) dibutyl dithiocarbamate	Oxidants added to the DESI spray	[81]
2006	Cooks et al.	Reactive DESI	TATP	Complexes with alkali metal ions (Li^+, Na^+ and K^+)	[25]
2007	Basile et al.	DESI	Proteins	12 to 66 kDa proteins detected	[82]
2007	Basile et al.	DESI	Intact bacteria	DESI-mass spectra, in the 50–500 u mass range obtained from whole bacteria	[83]
2006	Clemmer, Cooks et al.	DESI	Proteins	Coupling of reduced pressure ion mobility spectrometry to DESI	[84]

(Continued)

TABLE 8.1. *(Continued)*

Year	Authors	Ionization Technique Used	Application	Comments	Reference
2005	Creaser et al.	DESI	Pharmaceuticals	Coupling of reduced pressure ion mobility spectrometry to DESI	[28]
2007	Creaser et al.	DESI	Tryptic peptides	Coupling of reduced pressure ion mobility spectrometry to DESI	[29]
2006	Anderson et al.	DESI	Standard peptides	Improved ion transport with flared capillary inlet	[85]
2006	Kostiainen, Cooks et al.	DESI	Pharmaceuticals and metabolites	Different surfaces and sprayers tested	[86]
2006	Kostiainen, Cooks et al.	DESI	Pharmaceuticals	Porous silicon and UTLC surfaces tested	[87]
2007	Lepage et al.	DESI	Chemical warfare agents	Direct desorption from solid-phase microextraction fibers	[88]
2006	Lepage et al.	DESI	Chemical warfare agents in office media	DESI and LC-MS compared	[89]
2007	Muddiman et al.	DESI	Non-proximal sampling	Air ejector used as DESI-MS interface	[90]
2007	Muddiman et al.	DESI	Rhodamine 6G	Evaluation of DESI detection limits	[91]
2006	Fernandez, Muddiman et al.	DESI	Standard proteins	Coupling of DESI with FTICR-MS	[92]
2007	Fernandez, Kazarian et al.	DESI	Counterfeit drugs	Combined DESI and FT-IR imaging	[93]
2007	Fernandez et al.	Reactive DESI	Counterfeit drugs	Alkylamines used to form specific non-covalent complexes	[15]
2005	Van Berkel et al.	DESI	FD&C dyes	TLC readout using DESI	[94]
2006	Van Berkel et al.	DESI	Rhodamine dyes separated on TLC plates	TLC readout using DESI	[95]
2007	Van Berkel et al.	DESI	Standard dyes	Imaging of planar surfaces	[96]
2005	Scrivens et al.	DESI	Pharmaceutical tablets, gels, and ointments	Accurate mass measurements	[32]
2006	Scrivens et al.	DESI	Pharmaceutical solid, liquid, and cream formulations	Polarity switching accurate mass DESI	[30]
2006	Scrivens et al.	DESI	Low molecular weight synthetic polymers	Accurate mass measurements for structural determinations	[97]
2007	Scrivens et al.	DESI	PEG polymers	Rapid characterization	[98]
2006	Hopfgartner et al.	DESI	Ecstasy tablets	Selectivity added by MS/MS in QqQ$_{LIT}$	[99]
2006	Rodriguez-Cruz	DESI	Controlled substances	Comparison to in-house spectral database	[100]

heated He stream (eq. 2), react with these clusters, forming protonated gas-phase molecules (ABH^+) (eq. 3).

$$He^*(g) + nH_2O(g) \rightarrow He(g) + (H_2O)_{n-1} H^+(g) + OH^-(g) \qquad (1)$$

$$AB(s, l) \xrightarrow{heat} AB(g) \qquad (2)$$

$$(H_2O)_{n-1} H^+(g) + AB(g) \rightarrow [AB + H]^+(g) + (n-1)H_2O(g) \qquad (3)$$

The predominance of this cluster-mediated mechanism seems to be consistent with most observations indicating that $(ABH)^+$ ions are the major species observed in DART-MS of polar molecules [5]. However, direct Penning ionization of the analyte (a reaction similar to eq. 1 but where He metastables directly react with the analyte rather than atmospheric water to produce $M^{+\cdot}$ ions) and charge exchange reactions can also occur depending on the distance between the ion source and the mass spectrometer. Changes in this distance alter the chemistry of the reactive ionization region, and thus the species observed in the mass spectrum.

DART has been applied to a variety of problems, including the direct (i.e., without sample preparation) screening of solid pharmaceutical samples [31], counterfeit drugs [34–37], writing inks [38], ink photoinitiators in food [39], pathogenic bacteria [40], flavors and fragrances [41], reaction products [42], self-assembled monolayers [43], and analytes separated by high performance thin layer chromatography [44]. Several other applications have been posted as on-line notes [45]. Although in some cases DART has been coupled to triple quadrupole mass spectrometers [41], most applications make use of orthogonal time-of-flight (TOF) [5, 31, 34–36, 38–40, 44], thus making DART-TOF-MS a powerful tool for the rapid screening of solid samples in the open air followed by analyte identification via accurate mass measurements and isotope ratio matching.

Several other discharge-based ambient ionization techniques have been reported in the literature. McEwen et al. reported a method named atmospheric-pressure solids analysis probe (ASAP), where a solid sample probe is introduced directly into the corona discharge region of an enclosed atmospheric pressure chemical ionization (APCI) ion source via a custom built sample port. In ASAP, the heated gas stream from an APCI probe desorbs the analytes. Ionization of the thermally induced vapors occurs by corona discharge under standard APCI conditions. ASAP has been applied to a variety of analytical problems, including biological tissue, pharmaceuticals, polymers, currency [11], and inhibitors of the ergosterol pathway [46].

Zhang et al. reported a new technique termed dielectric barrier discharge ionization (DBDI) [10, 47], which they employed for the direct analysis of hexahydro-1,3,5-trinitro-1,3,5-triazine (RDX), 2,4,6-trinitrotoluene (TNT), and pentaerythritol tetranitrate (PETN) from explosives-contaminated surfaces, and for amino acids. In DBDI, a needle–plate electrical discharge generates energetic species, such as electrons, and these species apparently launch the desorption and ionization of the analyte from a solid surface, also used as a dielectric barrier, where the sample is deposited. In negative ion mode, DBDI was observed to form the typical [TNT]⁻, [TNT − H]⁻, [RDX + NO₂]⁻, [PETN + ONO₂]⁻, and [RDX + ONO₂]⁻ anions. For amino acids in positive ion mode, the $[M + H]^+$ protonated molecules were observed.

Desorption atmospheric pressure chemical ionization (DAPCI), originally implemented using toluene as the reagent gas flowing through the annular gas flow of a modified DESI sprayer in which the spray capillary was replaced by a sharp stainless steel tip [8], has been shown to be more effective at ionizing compounds of moderate to low polarity than DESI [31]. In DAPCI, the sample is directly introduced in the corona discharge region generated between a sharp needle and the capillary inlet of a mass spectrometer. Qiao et al. recently demonstrated the feasibility of using ambient air as the chemical ionization reagent in DAPCI [48]. A related method, plasma-assisted desorption/ionization (PADI) [9], has also been recently reported. PADI is carried out by generating a non-thermal RF plasma into which the sample is placed directly, as in ASAP, DBDI and DAPCI.

8.4 AMBIENT GAS-, HEAT-, AND LASER-ASSISTED DESORPTION/IONIZATION TECHNIQUES: ND-EESI, ELDI, LAESI, AND MALDESI

Ambient ionization methods in which the processes of desorption and ionization are decoupled enable soft ionization of analyte species as well as ionization of neutral species desorbed from the sample surface. Extractive electrospray ionization, introduced by the Cooks group in 2006, uses a gas stream to desorb compounds from a solid or liquid surface generating a neutral aerosol mixture that is subsequently ionized by interactions with an electrospray solution [49]. Zenobi et al. recently demonstrated the rapid profiling of complex biological samples such as frozen meat, vegetation, and human skin samples using neutral desorption extractive electrospray ionization (ND-EESI) [7] on a hybrid quadrupole-time-of-flight (Q-TOF) mass spectrometer. Mass accuracies of 10 ppm or better with detection limits as low as 10 femtograms histamine/cm^2 were achieved using this ionization technique with an analysis speed of 1–2 s.

Laser-assisted desorption/ionization techniques including electrospray-assisted laser desorption/ionization (ELDI), matrix-assisted laser desorption electrospray ionization (MALDESI), and laser ablation electrospray ionization (LAESI) combine laser desorption of analyte from the sample surface followed by electrospray ionization of the plume of particles, ions and neutrals generated. Shiea et al. used ELDI coupled to a Q-TOF to directly detect the active ingredient in drugs using UV irradiation to desorb methaqualone from a tablet followed by electrospray ionization [12]. MALDESI, which Muddiman et al. coupled to FT-ICR to obtain high resolution spectra from peptides and proteins, is similar to ELDI but uses an organic acid matrix to improve ion signal [13]. Vertes et al. used LAESI, which uses a mid-IR laser to irradiate water-rich samples to cause desorption, coupled to a TOF mass analyzer to monitor the excretion of the antihistamine fexofenadine in urine using accurate mass measurements [14].

8.5 CASE STUDY I: APPLICATION OF LIQUID CHROMATOGRAPHY ACCURATE-MASS MS TO THE FORENSIC EPIDEMIOLOGICAL CHARACTERIZATION OF COUNTERFEIT ARTESUNATE ANTIMALARIAL TABLETS

The most widely used definition of counterfeit drugs states that "A counterfeit drug is one which is deliberately and fraudulently mislabeled with respect to identity and/or source. Counterfeiting can apply to both branded and generic products and counterfeit products may include products with the correct ingredients or with the wrong ingredients, without active ingredients, with insufficient active ingredient" [50]. Drug counterfeiting is a widespread problem, as demonstrated by the many cases that have appeared in the news in the last decade [36, 37].

Antimalarials, among the most widely taken drugs in tropical countries, seem to have been particularly targeted by counterfeiters [51, 52]. Of the 12 major antimalarial drugs used in the world today, there are recent reports of 8 being counterfeited [36]. The control of malaria, predominantly dependent on effective antimalarial drugs and bed nets, has been severely hampered by a widespread increase in the prevalence of drug-resistant malaria parasites [53]. Artesunate, an artemisinin derivative [54], is widely and increasingly used in the treatment of drug-resistant *Plasmodium falciparum* malaria in many southeast Asian and African countries, and is vital for the therapy of drug-resistant malaria [55]. In Asia, artesunate has become the target of an extremely sophisticated counterfeit drug trade which includes counterfeiting of the tablets and packaging. Although there are at least 10 different manufacturers of artesunate tablets in Asia, only one manufacturer, Guilin Pharmaceutical Co., Ltd. (Guilin, Guangxi, People's Republic of China), appears to have been targeted by counterfeiters as all counterfeit artesunate tablets identified thus far were sold in packaging that mimics that of Guilin.

Field surveys using colorimetric tests conducted in 1999–2000, and 2001–2002 in Cambodia, Lao PDR (Laos), Burma (Myanmar), the Thailand/Burma border, and Vietnam, demonstrated that 38% and 53%, respectively, of artesunate tablets contained no active ingredient [51, 56]. Following these surveys, we conducted a detailed investigation of a set of 34 representative counterfeit artesunate tablet samples using LC-TOF-MS [57]. A total of 23 samples (68%) were found to be counterfeit. Eight (35%) of the fake samples contained wrong active ingredients, which were identified as different erythromycins and paracetamol using accurate mass measurements. Raman spectroscopy identified calcium carbonate as an excipient in 9 (39%) of 23 fake samples. Multivariate unsupervised pattern recognition indicated two major clusters of artesunate counterfeits, those with counterfeit foil stickers and containing calcium carbonate, erythromycin, and paracetamol, and those with counterfeit holograms and containing starch but without evidence of erythromycin or paracetamol. One genuine sample collected in Cambodia was shown to contain substandard amounts of artesunate (21 ± 2 mg). Subsequent work using DART-TOF-MS and LC with UV detection, detected the presence of fake artesunate tablets with small amounts of active ingredients [35], possibly added in an attempt to fool the current colorimetric authenticity tests [58]. More recent work using DESI-MS and MS/MS, detected the presence of another counterfeit sample with traces of artesunate on its surface, but not in the bulk of the tablet [15]. These and other investigations have resulted in several arrests, but it is still unclear if the complete counterfeiting operation has been dismantled [59].

8.6   CASE STUDY II: SURVEY OF LOW QUALITY COMBINATION MEDICINES BY AMBIENT IONIZATION ACCURATE-MASS MS

Between July 2000 and January 2001, a Thai-, Burmese-, and Karen-speaking researcher dressed as a Burmese migrant worker visited 47 "kong cham" shops, in 16 villages astride the Thai-Myanmar border in Mae Sot District, Tak Province, Thailand and Mywaddy District, Myanmar, with the intent of surveying a wide-spread class of medicines, know as "yaa chud" in Thai, or literally "combination medicine." These are sold to patients in small plastic bags containing four or five tablets and capsules without prescription or medical assessment. The Thai-Myanmar border is a region where multi-drug resistant *Plasmodium falciparum* malaria is a severe public health problem [60] and thus these drugs are very frequently used as a first line of treatment for fever and malaria. A total of 58 yaa chud bags (~3.1 tablets or capsules per bag) were screened using DART-TOF-MS with the intent of identifying, by accurate mass measurements, the active ingredients present in these low quality pharmaceuticals. This is of critical importance, as it was not known, for example, if yaa chud may contain drugs contraindicated in pregnancy or childhood.

Time-of-flight mass spectrometric measurements were performed with a JMS-100TLC (AccuTOF™) orthogonal TOF mass spectrometer (JEOL, USA, Peabody MA) equipped with a DART ionization source (IonSense Inc., Saugus MA). The DART ion source and the AccuTOF mass spectrometer were operated in positive-ion mode. Helium gas (4 L min^{-1}) was introduced into the DART corona discharge chamber where a needle electrode was held at 3.5 kV. The first DART ion source electrode was held at 150 V, and the grid electrode at 250 V. The DART gas temperature was 275 °C. Samples were manually held with a pair of metal tweezers in front of the DART ion source for an average time of 20 s. The total acquisition time to screen the complete sample set was 142 minutes, which corresponded to 207 sample spectra and 179 PEG spectra. In the case of coated tablets, these were first broken in half, and the interior of the sample was then analyzed by DART. Capsules were opened, and the contents were sampled with the open end of a melting point capillary (for powders), or with tweezers (for granules). No memory effects were observed as long as care was taken not to contact the inlet orifice of the mass spectrometer with the tablet.

Mass drift correction was performed by placing a 1.5 mm o.d. × 90 mm long glass capillary dipped in neat poly(ethylene glycol) (PEG, average molecular weight 600) in front of the DART helium stream for 2–5 s, and obtaining a reference mass spectrum immediately after each sample run. Mass spectral data centroiding, drift correction, and background subtraction were performed using the built-in mass spectrometer software (MassCenter, version 1.3). For identification purposes, these spectra were first exported into MS Excel, and searched, via a system of macros, against an in-house library of protonated molecules derived from drugs in the Model List of Essential Drugs published by the World Health Organization (WHO) [61]. A match was considered positive if the difference between the experimental and theoretical m/z values was less than 5 mmu. Spectra were continuously acquired over 10–20 minute runs, during which time samples were sequentially placed within the DART ionization region. During data post-processing and analysis, a mass spectrum of each unknown was obtained by averaging the mass spectral data over the time interval corresponding to the exposure of the corresponding sample to the DART gas.

Figure 8.1 (top left panel) shows a typical bag of yaa-chud containing five different tablets and capsules. The DART–TOF mass spectrum obtained for each tablet is shown in Figure 8.1(a–e). Observed DART spectra for yaa-chud samples contained only a few prominent peaks, as expected for simple pharmaceutical preparations. This spectral simplicity is also due to the absence of Na^+, K^+, and NH_4^+ adducts [34]. For all DART spectra, the vast majority of the observed signals corresponded to either $[M + H]^+$ or $[2M + H]^+$ ions, together with their corresponding isotopomers.

Due to the primitive packaging of the yaa-chud samples, cross contamination between tablets was common. Contaminants embedded on the sample surface produced signals corresponding to active ingredients from other tablets in the package (Figure 8.1(a–e)). If cross contamination between samples in a given package was suspected, tablets were broken in half, and the internal surface was also subjected to DART-MS. The active ingredients were thus assigned based on the common peaks observed for both, the external tablet surface, and the internal surface of the broken tablet. Of 182 samples tested, 165 (91%) were definitively identified using accurate mass measurements and isotope ratio matching using the DART data. The following were the most common compounds identified in the yaa chud collection: acetaminophen (33.5%), chlorpheniramine (18.1%), and chloroquine (16.5%). All DART findings were also verified by dissolving a small amount of the solid sample in 50:50:0.1 (v/v) methanol:water:acetic acid, and analyzing this solution by electrospray ionization (ESI) in continuous infusion mode. Complete agreement was found between ESI and DART-TOF-MS results. A complete compilation of sample photos, together with the compounds identified by ESI and DART for the entire yaa chud collection can be downloaded from the following URL [62].

Following the analysis of the yaa chud collection, we evaluated different approaches for performing accurate mass measurements via DART-TOF-MS. The effectiveness of any methodology for establishing the elemental composition of an analyte from its experimental accurate mass depends on properly establishing the mass error limits used to identify candidate elemental formulas. Although a single mass calibration check per day or per week is generally sufficient to ensure the accuracy of m/z measurements for most TOF mass analyzers, it is a good practice to incorporate a mass drift correction when performing long analyses, such as for long HPLC runs, or if temporal shifts in the local environment are expected. As

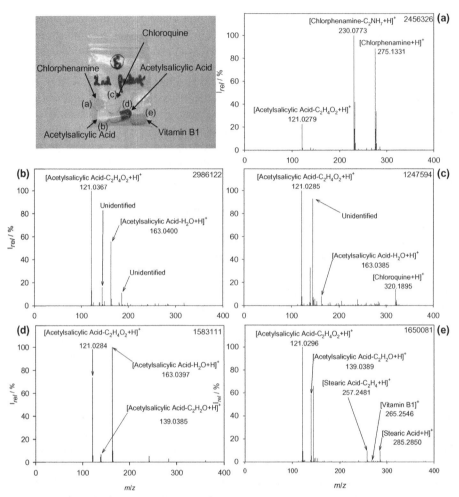

Figure 8.1. Accurate mass DART-TOF-MS mass spectra obtained from tablets and capsules found in a representative yaa-chud bag (shown in top left panel). Peaks were assigned if a match was found within a 5 mmu tolerance.

mentioned earlier, each sample was followed by a PEG standard in order to correct the polynomial coefficients used in the mass calibration file and compensate for drifts in the m/z scale. When screening sample collections containing in excess of 182 samples, such as in this survey, this approach can be time consuming, as it involves repeating the mass drift correction procedure 182 times.

Any drift in mass calibration during a DART run is usually due to room temperature changes, since both the power supply output and the length of the flight tube show thermal dependency [63]. Evaluating the extent to which these changes affect the observed mass accuracy is thus important in deciding the limits for matching candidates in our database, and for the calculation of elemental compositions using software tools. If the mass accuracy threshold is set too low, then it is possible to miss the actual compound. If the mass accuracy threshold is set too high, then too many matches can be obtained. In order to estimate the mass accuracies obtainable by alternative DART-TOF-MS data processing strategies, we tested two approaches—a single mass drift correction and a time-dependent mass drift correction.

In the single mass drift correction approach, multiple peaks (>5) within the PEG spectrum collected at the beginning of the DART run are compared to a reference file and are used to create a single calibration file. This calibration file is then applied to all sample spectra collected within the run. The primary benefit of this approach is the reduction in time required to collect and analyze the data. However, the advantage of increased throughput using this method may be offset by lowered mass accuracies. To explore this further, a second mass drift correction approach was performed. For the time-dependent mass drift correction, an individual calibration file is created for each PEG spectrum collected during the run. The calibration file for the PEG spectrum collected immediately after a given sample spectrum is then applied as a correction for that sample, thus, each sample spectrum is corrected using the PEG reference spectrum most closely associated to the sample spectrum in time.

Figure 8.2 shows the mass accuracies observed for the drugs identified in the yaa-chud samples for three different DART-TOF-MS runs. The average mass accuracy for each DART run was 3.3 ± 1.7 ppm, 5.7 ± 5.1 ppm, and 5.3 ± 2.7 ppm ($\bar{x} \pm s$) using the single mass drift correction, and 2.2 ± 2.1 ppm, 2.4 ± 1.7 ppm, and 3.0 ± 2.5 ppm ($\bar{x} \pm s$) using the time-dependent mass drift correction. Significant statistical differences in the observed mass accuracy were observed for individual runs between the two mass drift correction approaches used. The differences in observed mass accuracy were greatest for samples analyzed at the end of the runs emphasizing the importance of using the more rigorous time-dependent mass drift correction approach for analyses with long run times. The average mass accuracy of each run was better using the time-dependent drift correction for each sample; however, the throughput of this approach is lower. For example, using the average time it took us to collect a spectrum (22.1 seconds), 4.1 minutes would be required to collect data for 10 samples (and 1 PEG spectrum) analyzed in a single run using the single mass drift correction approach. Using the time-dependent approach requires the analysis of an additional nine reference spectra which adds 3.4 minutes to the collection time. The data analysis time would also be increased using this method. The decision as to whether throughput or accuracy is the determining factor for which of these approaches to apply is dictated by the requirements of the experiment and the use of shorter run times would appear to negate this conflict entirely. The mean mass accuracy observed for all PEG standards was 3.8 ± 3.2 ppm ($\bar{x} \pm s$), which is comparable with the manufacturer's specification of 5 ppm for electrospray TOF accurate mass measurements. Seventy-three percent of the PEG standard measurements had mass accuracies of 5 ppm or better, and 94.5% equal or better than 10 ppm. Similar performance has been observed for TOF analyzers coupled with electrospray ion sources [64, 65], indicating that the DART ion source did not introduce additional variables which negatively influenced mass accuracy.

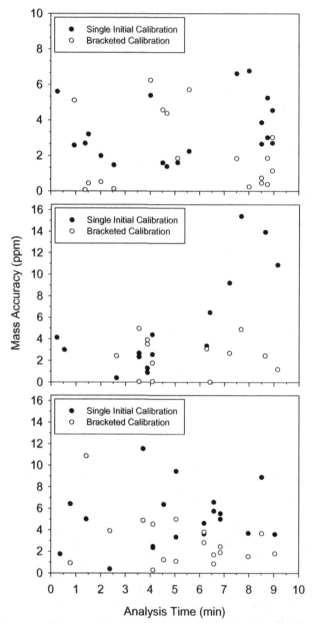

Figure 8.2. Observed mass accuracy for drugs analyzed during three independent DART-TOF-MS runs. Open circles show mass accuracies obtained for each individually-corrected sample. Full circles show mass accuracies observed when a single initial mass drift correction was employed for the whole run.

8.7 CONCLUSIONS

The benefits of accurate mass spectrometry, namely the ability to perform accurate, sensitive, non-targeted identification and quantitation of analytes, were presented in the context of two case studies. The use of LC-MS, the method of choice for the routine analysis of samples in complex matrices or containing multiple components, was described for the characterization of a set of antimalarial pharmaceuticals. The accurate mass measurements obtained were used to identify the components of these drugs and classify them according to their geographical origin. An overview of several different ambient ionization methods was also presented with an effort to describe the applicability of each method. A detailed study of combination drug packets screened by DART-MS demonstrated the ability of this latest generation of ionization methods to achieve high throughput accurate mass identifications of 91% of 182 samples analyzed during this survey. Two approaches to performing mass calibrations were explored in an effort to identify the factors which most influence the quality of mass spectral data obtained during accurate mass experiments. The results emphasize the importance of using a mass correction methodology that addresses the needs of the experiment being executed.

ACKNOWLEDGMENTS

The authors are thankful for financial support from the US Pharmacopeia to LN, an NSF CAREER award to FMF, and by WPRO/WHO. We are also extremely grateful for our on-going collaboration with Dr. Paul Newton and Dr. Michael Green on the epidemiological forensics of counterfeit antimalarial drugs. The collection of antimalarial samples is funded by the Wellcome Trust of Great Britain as part of the Wellcome Trust-South East Asian Oxford Tropical Medicine Research Collaboration.

REFERENCES

1. Cotter, R.J., The new time-of-flight mass spectrometry. *Anal. Chem.*, **1999**, 71, 445A–451A.
2. Cooks, R.G.; Ouyang, Z.; Takats, Z.; Wiseman, J.M., Ambient mass spectrometry. *Science*, **2006**, 311, 1566–1570.
3. Chen, H.; Talaty, N.N.; Takats, Z.; Cooks, R.G., Desorption electrospray ionization mass spectrometry for high-throughput analysis of pharmaceutical samples in the ambient environment. *Anal. Chem.*, **2005**, 77, 6915–6927.
4. Takats, Z.; Wiseman, J.M.; Gologan, B.; Cooks, R.G., Mass spectrometry sampling under ambient conditions with desorption electrospray ionization. *Science*, **2004**, 306, 471–473.
5. Cody, R.; Laramee, J.; Durst, H., Versatile new ion source for the analysis of materials in open air under ambient conditions. *Anal. Chem.*, **2005**, 77, 2297–2302.
6. Haddad, R.; Sparrapan, R.; Eberlin, M.N., Desorption sonic spray ionization for (high) voltage-free ambient mass spectrometry. *Rapid Commun. Mass Spectrom.*, **2006**, 20, 2901–2905.
7. Chen, H.; Wortmann, A.; Zenobi, R., Neutral desorption sampling coupled to extractive electrospray ionization mass spectrometry for rapid differentiation of biosamples by metabolomic fingerprinting. *J. Mass Spectrom.*, **2007**, 42, 1123–1135.

8. Takats, Z.; Cotte-Rodriguez, I.; Talaty, N.; Chen, H.W.; Cooks, R.G., Direct, trace level detection of explosives on ambient surfaces by desorption electrospray ionization mass spectrometry. *Chem. Commun.*, **2005**, 15, 1950–1952.

9. Ratcliffe, L.V.; Rutten, F.J.M.; Barrett, D.A.; Whitmore, T.; Seymour, D.; Greenwood, C.; Aranda-Gonzalvo, Y.; Robinson, S.; McCoustrat, M., Surface analysis under ambient conditions using plasma-assisted desorption/ionization mass spectrometry. *Anal. Chem.*, **2007**, 79, 6094–6101.

10. Na, N.; Zhao, M.; Zhang, S.; Yang, C.L.; Zhang, X., Development of a dielectric barrier discharge ion source for ambient mass spectrometry. *J. Am. Soc. Mass Spectrom.*, **2007**, 18, 1859–1862.

11. McEwen, C.N.; McKay, R.G.; Larsen, B.S., Analysis of solids, liquids, and biological tissues using solids probe introduction at atmospheric pressure on commercial LC/MS instruments. *Anal. Chem.*, **2005**, 77, 7826–7831.

12. Shiea, J.; Huang, M.Z.; HSu, H.J.; Lee, C.Y.; Yuan, C.H.; Beech, I.; Sunner, J., Electrospray-assisted laser desorption/ionization mass spectrometry for direct ambient analysis of solids. *Rapid Commun. Mass Spectrom.*, **2005**, 19, 3701–3704.

13. Sampson, J.S.; Hawkridge, A.M.; Muddiman, D.C., Generation and detection of multiply-charged peptides and proteins by matrix-assisted laser desorption electrospray ionization (MALDESI) fourier transform ion cyclotron resonance mass spectrometry. *J. Am. Soc. Mass Spectrom.*, **2006**, 17, 1712–1716.

14. Nemes, P.; Vertes, A., Laser ablation electrospray ionization for atmospheric pressure, in vivo, and imaging mass spectrometry. *Anal. Chem.*, **2007**, 79, 8098–8106.

15. Nyadong, L.; Green, M.; De Jesus, V.; Newton, P.N.; Fernandez, F.M., Reactive Desorption electrospray ionization linear ion trap mass spectrometry of latest-generation counterfeit antimalarials via non-covalent complex formation. *Anal. Chem.*, **2007**, 79, 2150–2157.

16. Takats, Z.; Wiseman, J.M.; Cooks, R.G., Ambient mass spectrometry using desorption electrospray ionization (DESI): instrumentation, mechanisms and applications in forensics, chemistry, and biology. *J. Mass Spectrom.*, **2005**, 40, 1261–1275.

17. Vincenti, M.; Cooks, R.G., Desorption due to charge-exchange in low-energy collisions of organofluorine ions at solid-surfaces. *Org. Mass Spectrom.*, **1988**, 23, 317–326.

18. Cooks, R.G.; Ast, T.; Mabud, A., Collisions of polyatomic ions with surfaces. *Int. J. Mass Spectrom. Ion Processes*, **1990**, 100, 209–265.

19. Venter, A.; Sojka, P.E.; Cooks, R.G., Droplet dynamics and ionization mechanisms in desorption electrospray ionization mass spectrometry. *Anal. Chem.*, **2006**, 78, 8549–8555.

20. Venter, A.; Sojka, P.E.; Cooks, R.G., *Kinematic Investigation of Charged Droplets for Mechanistic Considerations of Desorption Electrospray Ionization (DESI)*, 54th ASMS Conference on Mass Spectrometry, Seattle, WA, **2006**.

21. Costa, A.B.; Cooks, R.G., Simulation of atmospheric transport and droplet–thin film collisions in desorption electrospray ionization. *Chem. Commun.*, **2007**, 3915–3917.

22. Yang, P.X.; Cooks, R.G.; Ouyang, Z.; Hawkridge, A.M.; Muddiman, D.C., Gentle protein ionization assisted by high-velocity gas flow. *Anal. Chem.*, **2005**, 77, 6174–6183.

23. Wiseman, J.M.; Takats, Z.; Gologan, B.; Davisson, V.J.; Cooks, R.G., Direct characterization of enzyme-substrate complexes by using electrosonic spray ionization mass spectrometry. *Angew. Chem., Int. Ed.*, **2005**, 44, 913–916.

24. Hirabayashi, A.; Sakairi, M.; Koizumi, H., Sonic spray mass-spectrometry. *Anal. Chem.*, **1995**, 67, 2878–2882.

25. Cotte-Rodriguez, I.; Chen, H.; Cooks, R.G., Rapid trace detection of triacetone triperoxide (TATP) by complexation reactions during desorption electrospray ionization. *Chem. Commun.*, **2006**, 953–955.

26. Cotte-Rodriguez, I.; Takats, Z.; Talaty, N.; Chen, H.; Cooks, R.G., Desorption electrospray ionization of explosives on surfaces: sensitivity and selectivity enhancement by reactive desorption electrospray ionization. *Anal. Chem.*, **2005**, 77, 6755–6764.

27. Hu, Q.Z.; Talaty, N.; Noll, R.J.; Cooks, R.G., Desorption electrospray ionization using an orbitrap mass spectrometer: exact mass measurements on drugs and peptides. *Rapid Commun. Mass Spectrom.*, **2006**, 20, 3403–3408.

28. Weston, D.J.; Bateman, R.; Wilson, I.D.; Wood, T.R.; Creaser, C.S., Direct analysis of pharmaceutical drug formulations using ion mobility spectrometry/quadrupole-time-of-flight mass spectrometry combined with desorption electrospray ionization. *Anal. Chem.*, **2005**, 77, 7572–7580.

29. Kaur-Atwal, G.; Weston, D.J.; Green, P.S.; Crosland, S.; Bonner, P.L.R.; Creaser, C.S., Analysis of tryptic peptides using desorption electrospray ionisation combined with ion mobility spectrometry/mass spectrometry. *Rapid Commun. Mass Spectrom.*, **2007**, 21, 1131–1138.

30. Williams, J.P.; Lock, R.; Patel, V.J.; Scrivens, J.H., Polarity switching accurate mass measurement of pharmaceutical samples using desorption electrospray ionization and a dual ion source interfaced to an orthogonal acceleration time-of-flight mass spectrometer. *Anal. Chem.*, **2006**, 78, 7440–7445.

31. Williams, J.P.; Patel, V.J.; Holland, R.; Scrivens, J.H., The use of recently described ionisation techniques for the rapid analysis of some common drugs and samples of biological origin. *Rapid Commun. Mass Spectrom.*, **2006**, 20, 1447–1456.

32. Williams, J.P.; Scrivens, J.H., Rapid accurate mass desorption electrospray ionisation tandem mass spectrometry of pharmaceutical samples. *Rapid Commun. Mass Spectrom.*, **2005**, 19, 3643–3650.

33. Penning, F.M., Ionization by metastable atoms. *Naturwissenschaften*, **1927**, 15, 818.

34. Fernandez, F.M.; Cody, R.B.; Green, M.; Hampton, C.Y.; McGready, R.; Sengaloundeth, S.; White, N.J.; Newton, P.N., Characterization of solid counterfeit drug samples by desorption electrospray ionization and direct-analysis-in-real-time coupled to time-of-flight mass spectrometry. *Chem. Med. Chem.*, **2006**, 1, 702–705.

35. Newton, P.N.; McGready, R.; Fernandez, F.M.; Green, M.D.; Sunjio, M.; Bruneton, C.; Phanouvong, S.; Millet, P.; C.J., W.; Talisuna, A.O.; Proux, S.; Christophel, E.M.; Malenga, G.; Singhasivanon, P.; Bojang, K.; Kaur, H.; Palmer, K.; Day, N.P.J.; Greenwood, B.M.; Nosten, F.; White, N.J., Manslaughter by fake artesunate in Asia—will Africa be next? *PLoS Medicine*, **2006**, 3, e197.

36. Newton, P.N.; Green, M.; Fernandez, F.M.; Day, N.P.J.; White, N.J., Counterfeit anti-infective drugs. *Lancet Infect. Dis.*, **2006**, 6, 602–612.

37. Fernandez, F.M.; Newton, P.N.; Green, M., Prevalence and detection of counterfeit pharmaceuticals: a mini review. *Ind. Eng. Chem. Res.*, **2008**, 47, 585–590.

38. Jones, R.W.; Cody, R.B.; McClelland, J.F., Differentiating writing inks using direct analysis in real time mass spectrometry. *J. Forensic Sci.*, **2006**, 51, 915–918.

39. Morlock, G.; Schwack, W., Determination of isopropylthioxanthone (ITX) in milk, yoghurt and fat by HPTLC-FLD, HPTLC-ESI/MS and HPTLC-DART/MS. *Anal. Bioanal. Chem.*, **2006**, 385, 586–595.

40. Pierce, C.Y.; Barr, J.R.; Cody, R.B.; Massung, R.; Woolfitt, A.; Moura, H.; Thompson, H.A.; Fernandez, F.M., Ambient generation of fatty acid methyl ester ions from bacterial whole cells by direct analysis in real time (DART) mass spectrometry. *Chem. Commun.*, **2007**, 807–809.

41. Haefliger, O.P.; Jeckelmann, N., Direct mass spectrometric analysis of flavors and fragrances in real applications using DART. *Rapid Commun. Mass Spectrom.*, **2007**, 21, 1361–1366.

42. Petucci, C.; Diffendal, J.; Kaufman, D.; Mekonnen, B.; Terefenko, G.; Musselman, B., Direct analysis in real time for reaction monitoring in drug discovery. *Anal. Chem.*, **2007**, 79, 5064–5070.

43. Kpegba, K.; Spadaro, T.; Cody, R.B.; Nesnas, N.; Olson, J.A., Analysis of self-assembled monolayers on gold surfaces using direct analysis in real time mass spectrometry. *Anal. Chem.*, **2007**, 79, 5479–5483.

44. Morlock, G.; Ueda, Y., New coupling of planar chromatography with direct analysis in real time mass spectrometry. *J. Chromatogr. A*, **2007**, 1143, 243–251.

45. IonSense http://www.ionsense.com/applications.php.

46. McEwen, C.N.; Gutteridge, S., Analysis of the inhibition of the ergosterol pathway in fungi using the atmospheric solids analysis probe (ASAP) method. *J. Am. Soc. Mass Spectrom.*, **2007**, 18, 1274–1278.

47. Na, N.; Zhao, M.; Zhang, S.; Yang, C.L.; Zhang, X., Development of a dielectric barrier discharge ion source for ambient mass spectrometry. *J. Am. Soc. Mass Spectrom.*, **2007**, 18, 1859–1862.

48. Chen, H.; Zheng, J.; Zhang, X.; Luo, M.; Wang, Z.; Qiao, X., Surface desorption atmospheric pressure chemical ionization mass spectrometry for direct ambient sample analysis without toxic chemical contamination. *J. Mass Spectrom.*, **2007**, 42, 1045–1056.

49. Chen, H.; Venter, A.; Cooks, R.G., Extractive electrospray ionization for direct analysis of undiluted urine, milk and other complex mixtures without sample preparation. *Chem. Commun.*, **2006**, 2042–2044.

50. WHO Counterfeit Drugs. Guidelines for the Development of Measures to Combat Counterfeit Drugs. 1999, 99.1, http://www.who.int/medicines/publications/counterfeitguidelines/en/.

51. Newton, P.; Proux, S.; Green, M.; Smithuis, F.; Rozendaal, J.; Prakongpan, S.; Chotivanich, K.; Mayxay, M.; Looareesuwan, S.; Farrar, J.; Nosten, F.; White, N.J., Fake artesunate in Southeast Asia. *The Lancet*, **2001**, 357, 1948–1950.

52. Basco, L.K., Molecular epidemiology of malaria in Cameroon. XIX. Quality of antimalarial drugs used for self-medication. *Am. J. Trop. Med. Hyg.*, **2004**, 70, 245–250.

53. White, N.J., Antimalarial drug resistance. *J. Clin. Invest.*, **2004**, 113, 1084–1092.

54. Meshnick, S.R., Artemisinin: mechanisms of action, resistance and toxicity. *Int. J. Parasitol.*, **2002**, 32, 1655–1660.

55. van Agtmael, M.A.; Eggelte, T.A.; van Boxtel, C.J., Artemisinin drugs in the treatment of malaria: from medicinal herb to registered medication. *Trends Pharmacol. Sci.*, **1999**, 20, 199–205.

56. Dondorp, A.M.; Newton, P.N.; Mayxay, M.; van Damme, W.; Smithuis, F.M.; Yeung, S.; Petit, A.; Lynam, A.J.; Johnson, A.; Hien, T.T.; McGready, R.; Farrar, J.J.; Looareesuwan, S.; Day, N.P.J.; Green, M.D.; White, N.J., Fake antimalarials in Southeast Asia are a major impediment to malaria control: multinational cross-sectional survey on the prevalence of fake antimalarials. *Trop. Med. Int. Health*, **2004**, 9, 1241–1246.

57. Alter Hall, K.; Newton, P.N.; Green, M.D.; De Veij, M.; Vandenabeele, P.; Pizzanelli, D.; Mayxay, M.; Dondorp, A.; White, N.J.; Fernandez, F.M., Characterization of counterfeit artesunate antimalarial tablets from SE Asia. *Am. J. Trop. Med. Hyg.*, **2006**, 75, 804–811.

58. Green, M.D.; Nettey, H.; Villalba-Rojas, O.; Pamanivong, C.; Khounsaknalath, L.; Grande Ortiz, M.; Newton, P.N.; Fernandez, F.M.; Vongsack, L.; Manolin, O., Use of refractometry and colorimetry as field methods to rapidly assess antimalarial drug quality. *J. Pharm. Biomed. Anal.*, **2007**, 43, 105–110.

59. Newton, P.N.; Fernandez, F.M.; Plancon-Lecadre, A.; Mildenhall, D.; Green, M.D.; Ziyong, L.; Christophel, E.M.; Phanouvong, S.; Howells, S.; MacIntosh, E.; Laurin, P.; Blum, N.; Hampton, C.Y.; Faure, K.; Nyadong, L.; Soong, C.W.R.; Santoso, B.; Zhiguang, W.; Newton, J.; Palmer, K., A collaborative epidemiological investigation into the criminal fake artesunate trade in SE Asia. *PLoS Medicine*, **2008**, 5, e32.

60. Nosten, F.; van Vugt, M.; Price, R.; Luxemburger, C.; Thway, K.L.; Brockman, A.; McGready, R.; ter Kuile, F.; Looareesuwan, S.; White, N.J., Effects of artesunate-mefloquine combination on incidence of *plasmodium falciparum* malaria and mefloquine resistance in Western Thailand: a prospective study. *Lancet*, **2000**, 356, 297–302.

61. WHO Who Model List of Essential Medicines. http://www.who.int/medicines/publications/essentialmedicines/en/.

62. Hampton, C.Y.; Leung, H.; Fernandez, F.M., Yaa Chud Compilation. http://web.chemistry.gatech.edu/~fernandez/Yaa_chud_compilation.pdf, **2007**.

63. Chernushevich, I.V.; Loboda, A.V.; Thomson, B.A., An introduction to quadrupole-time-of-flight mass spectrometry. *J. Mass Spectrom.*, **2001**, 36, 849–865.

64. Blom, K.F., Estimating the precision of exact mass measurements on an orthogonal time-of-flight mass spectrometer. *Anal. Chem.*, **2001**, 73, 715–719.

65. Wolff, J.C.; Fuentes, T.R.; Taylor, J., Investigations into the accuracy and precision obtainable on accurate mass measurements on a quadrupole orthogonal acceleration time-of-flight mass spectrometer using liquid chromatography as sample introduction. *Rapid Commun. Mass Spectrom.*, **2003**, 17, 1216–1219.

66. Wiseman, J.M.; Ifa, D.R.; Song, Q.; Cooks, R.G., Tissue imaging at atmospheric pressure using desorption electrospray ionization (DESI) mass spectrometry *Angew. Chem. Int. Ed.*, **2006**, 45, 7188–7192.

67. Venter, A.; Cooks, R.G., Desorption electrospray ionization in a small pressure-tight enclosure. *Anal. Chem.*, **2007**, 79, 6398–6403.
68. Cotte-Rodriguez, I.; Mulligan, C.C.; Cooks, R.G., Non-proximate detection of small and large molecules by desorption electrospray ionization and desorption atmospheric pressure chemical ionization mass spectrometry: instrumentation and applications in forensics, chemistry, and biology. *Anal. Chem.*, **2007**, 79, 7069–7077.
69. Pan, Z.Z.; Gu, H.W.; Talaty, N.; Chen, H.W.; Shanaiah, N.; Hainline, B.E.; Cooks, R.G.; Raftery, D., Principal component analysis of urine metabolites detected by NMR and DESI-MS in patients with inborn errors of metabolism. *Anal. Bioanal. Chem.*, **2007**, 387, 539–549.
70. Mulligan, C.C.; Talaty, N.; Cooks, R.G., Desorption electrospray ionization with a portable mass spectrometer: in situ analysis of ambient surfaces. *Chem. Commun.*, **2006**, 1709–1711.
71. Nefliu, M.; Venter, A.; Cooks, R.G., Desorption electrospray ionization and electrosonic spray ionization for solid- and solution-phase analysis of industrial polymers. *Chem. Commun.*, **2006**, 888–890.
72. Wiseman, J.M.; Puolitaival, S.M.; Takats, Z.; Cooks, R.G.; Caprioli, R.M., Mass spectrometric profiling of intact biological tissue by using desorption electrospray ionization. *Angew. Chem., Int. Ed.*, **2005**, 44, 7094–7097.
73. Chen, H.; Pan, Z.; Talaty, N.; Raftery, D.; Cooks, R.G., Combining desorption electrospray ionization mass spectrometry and nuclear magnetic resonance for differential metabolomics without sample preparation. *Rapid Commun. Mass Spectrom.*, **2006**, 20, 1577–1584.
74. Song, Y.; Talaty, N.; Tao, A.W.; Pan, Z.; Cooks, R.G., Rapid ambient mass spectrometric profiling of intact, untreated bacteria using desorption electrospray ionization. *Chem. Commun.*, **2007**, 61–63.
75. Cotte-Rodriguez, I.; Cooks, R.G., Non-proximate detection of explosives and chemical warfare agent simulants by desorption electrospray ionization mass spectrometry. *Chem. Commun.*, **2006**, 2968–2970.
76. Ifa, D.R.; Gumaelius, L.M.; Eberlin, L.S.; Manicke, N.E.; Cooks, R.G., Forensic analysis of inks by imaging desorption electrospray ionization (DESI) mass spectrometry. *Analyst*, **2007**, 132, 461–467.
77. Kauppila, T.J.; Talaty, N.; Kuuranne, T.; Kotiaho, T.; Kostiainen, R.; Cooks, R.G., Rapid analysis of metabolites and drugs of abuse from urine samples by desorption electrospray ionization-mass spectrometry. *Analyst*, **2007**, 132, 868–875.
78. Justes, D.R.; Talaty, N.; Cotte-Rodriguez, I.; Cooks, R.G., Detection of explosives on skin using ambient ionization mass spectrometry. *Chem. Commun.*, **2007**, 2142–2144.
79. Chen, H.; Cotte-Rodriguez, I.; Cooks, R.G., Cis-diol functional group recognition by reactive desorption electrospray ionization (DESI). *Chem. Commun.*, **2006**, 597–599.
80. Song, Y.; Cooks, R.G., Reactive desorption electrospray ionization for selective detection of the hydrolysis products of phosphonate esters. *J. Mass Spectrom.*, **2007**, 42, 1086–1092.
81. Nefliu, M.; Cooks, R.G.; Moore, C., Enhanced desorption ionization using oxidizing electrosprays. *J. Am. Soc. Mass Spectrom.*, **2006**, 17, 1091–1095.
82. Shin, Y.S.; Drolet, B.; Mayer, R.; Dolence, K.; Basile, F., Desorption electrospray ionization-mass spectrometry of proteins. *Anal. Chem.*, **2007**, 79, 3514–3518.
83. Meetani, M.A.; Shin, Y.S.; Zhang, S.; Mayer, R.; Basile, F., Desorption electrospray ionization mass spectrometry of intact bacteria. *J. Mass Spectrom.*, **2007**, 42, 1186–1193.
84. Myung, S.; Wiseman, J.M.; Valentine, S.J.; Takats, Z.; Cooks, R.G.; Clemmer, D.E., Coupling desorption electrospray ionization with ion mobility/mass spectrometry for analysis of protein structure: evidence for desorption of folded and denatured states. *J. Phys. Chem. B*, **2006**, 110, 5045–5051.
85. Wu, S.; Zhang, K.; Kaiser, N.K.; Bruce, J.E.; Prior, D.C.; Anderson, G.A., Incorporation of a flared inlet capillary tube on a Fourier transform ion cyclotron resonance mass spectrometer. *J. Am. Soc. Mass Spectrom.*, **2006**, 17, 772–779.
86. Kauppila, T.J.; Wiseman, J.M.; Ketola, R.A.; Kotiaho, T.; Cooks, R.G.; Kostiainen, R., Desorption electrospray ionization mass spectrometry for the analysis of pharmaceuticals and metabolites. *Rapid Commun. Mass Spectrom.*, **2006**, 20, 387–392.

87. Kauppila, T.J.; Talaty, N.; Salo, P.K.; Kotiah, T.; Kostiainen, R.; Cooks, R.G., New surfaces for desorption electrospray ionization mass spectrometry: porous silicon and ultra-thin layer chromatography plates. *Rapid Commun. Mass Spectrom.*, **2006**, 20, 2143–2150.

88. D'Agostino, P.A.; Chenier, C.L.; Hancock, J.R.; Lepage, C.R.J., Desorption electrospray ionisation mass spectrometric analysis of chemical warfare agents from solid-phase microextraction fibers. *Rapid Commun. Mass Spectrom.*, **2007**, 21, 543–549.

89. D'Agostino, P.A.; Hancock, J.R.; Chenier, C.L.; Lepage, C.R.J., Liquid chromatography electrospray tandem mass spectrometric and desorption electrospray ionization tandem mass spectrometric analysis of chemical warfare agents in office media typically collected during a forensic investigation. *J. Chromatogr. A*, **2006**, 1110, 86–94.

90. Dixon, R.B.; Bereman, M.S.; Muddiman, D.C.; Hawkridge, A.M., Remote mass spectrometric sampling of electrospray- and desorption electrospray-generated ions using an air ejector. *J. Am. Soc. Mass Spectrom.*, **2007**, 18, 1844–1847.

91. Bereman, M.S.; Muddiman, D.C., Detection of attomole amounts of analyte by desorption electrospray ionization mass spectrometry (DESI-MS) determined using fluorescence spectroscopy. *J. Am. Soc. Mass Spectrom.*, **2007**, 18, 1093–1096.

92. Bereman, M.S.; Nyadong, L.; Fernandez, F.M.; Muddiman, D.C., Direct high resolution peptide and protein analysis by desorption electrospray ionization (DESI) Fourier transform ion cyclotron resonance mass spectrometry. *Rapid Commun. Mass Spectrom.*, **2006**, 20, 3409–3411.

93. Ricci, C.; Nyadong, L.; Fernandez, F.M.; Newton, P.N.; Kazarian, S., Combined Fourier transform infrared imaging and desorption electrospray ionization linear ion trap mass spectrometry for the analysis of counterfeit antimalarial tablets. *Anal. Bioanal. Chem.*, **2007**, 387, 551–559.

94. Van Berkel, G.J.; Ford, M.J.; Deibel, M.A., Thin-layer chromatography and mass spectrometry coupled using desorption electrospray ionization. *Anal. Chem.*, **2005**, 77, 1207–1215.

95. Van Berkel, G.J.; Kertesz, V., Automated sampling and imaging of analytes separated on thin-layer chromatography plates using desorption electrospray ionization mass spectrometry. *Anal. Chem.*, **2006**, 78, 4938–4944.

96. Pasilis, S.P.; Kertesz, V.; Van Berkel, G.J., Surface scanning analysis of planar arrays of analytes with desorption electrospray ionization-mass spectrometry. *Anal. Chem.*, **2007**, 79, 5956–5962.

97. Jackson, A.T.; Williams, J.P.; Scrivens, J.H., Desorption electrospray ionisation mass spectrometry and tandem mass spectrometry of low molecular weight synthetic polymers. *Rapid Commun. Mass Spectrom.*, **2006**, 20, 2717–2727.

98. Williams, J.P.; Hilton, G.R.; Thalassinos, K.; Jackson, A.T.; Scrivens, J.H., The rapid characterisation of poly(ethylene glycol) oligomers using desorption electrospray ionisation tandem mass spectrometry combined with novel product ion peak assignment software. *Rapid Commun. Mass Spectrom.*, **2007**, 21, 1693–1704.

99. Leuthold, L.A.; Mandscheff, J.F.; Fathi, M.; Giroud, C.; Augsburger, M.; Varesio, E.; Hopfgartner, G., Desorption electrospray ionization mass spectrometry: direct toxicological screening and analysis of illicit ecstasy tablets. *Rapid Commun. Mass Spectrom.*, **2006**, 20, 103–110.

100. Rodriguez-Cruz, S.E., Rapid analysis of controlled substances using desorption electrospray ionization mass spectrometry. *Rapid Commun. Mass Spectrom.*, **2006**, 20, 53–60.

CHAPTER 9

QUANTITATIVE ANALYSIS OF VETERINARY DRUG RESIDUES BY SUB 2-μm PARTICULATE HIGH-PERFORMANCE LIQUID CHROMATOGRAPHY COLUMNS AND TIME-OF-FLIGHT MASS SPECTROMETRY (UPLC-TOF)

Anton Kaufmann

Official Food Control Authority of the Canton of Zurich, Kantonales Labor Zürich, Zürich, Switzerland

A MULTI-RESIDUE, multi-class method for the quantification of antibiotic residues in different meat matrix is described. The method covers more than 100 different analytes and is based on high resolution (sub 2 μm particulate columns) and time of flight mass spectrometry. This technique produced sufficient selectivity and sensitivity for trace levels of antibiotics present in difficult food matrices. The advantage of this approach, as compared to the more commonly used tandem mass spectrometry, lies in the absence of dwell time issues and the labor intensive definition of retention time-based transition windows. Furthermore, qualitative analysis of non-target compounds for which no reference compounds are available is possible.

9.1 INTRODUCTION

9.1.1 MS As A Quantitative Technique

Mass spectrometry (MS) was originally developed to provide qualitative data like masses and corresponding elemental composition of inorganic and organic

Liquid Chromatography Time-of-Flight Mass Spectrometry: Principles, Tools, and Applications for Accurate Mass Analysis, Edited by Imma Ferrer and E. Michael Thurman
Copyright © 2009 John Wiley & Sons, Inc.

compounds. Problems which initially prevented the successful quantification of analytes were related to the limited ruggedness of the instrument, the control of ionization and the narrow dynamic range. Reproducible ionization of a trace analyte in the presence of an overwhelming number and concentration of endogenous compounds proved to be challenging. This was significantly improved by using some form of separation prior to MS e.g., chromatography. Interfacing gas chromatography (GC) to MS was realized one or two decades before liquid chromatography (LC) could be reliably connected to MS. But only the introduction of the electrospray interface (ESI) gave LC-MS the necessary impetus. However hyphenation added additional requirements to mass spectrometers. They have to be able to follow the fast changing concentration of an analyte peak eluting from the chromatographic column and to be capable of recording a large dynamic range.

Interface Related Problems   Often, a mass spectrometer does not give the same response intensity if an equal analyte concentration is present in a neat standard solution or in a matrix extract. Such behavior is generally related to the ionization efficiency being affected by other compounds present in the matrix but not in the neat standard solution. Depending on the monitored analyte and the interfering matrix compounds, signal suppression or signal enhancement can be observed. Such effects have been reported—to various extent—for all types of interfaces. There were reports that certain interfaces like atmospheric pressure chemical ionization (APCI) and atmospheric pressure photo ionization (APPI) are less affected by signal suppression than ESI, however, such statements often refer to one particular analytical application where one interface was more rugged than the other [1]. A general conclusion is more difficult to draw, since many more methods have been developed for ESI interfaces than any other interfaces. Consequently, strengths and weaknesses of ESI are much better known than of APCI and APPI.

Signal suppression in ESI is thought to occur if the escape of analyte ions from the shrinking eluent droplets is hindered by some physical or chemical effects. This might involve the presence of some surface active matrix compounds which predominantly occupy the surface of the droplets, hence hindering the evaporation of analyte ions. The formation of adducts (e.g., sodium, ammonium, chloride) as often observed in ESI, affects quantitation by producing additional mass peaks at the expense of $[M+H]^+$ or $[M-H]^-$ ions. The ratio of $[M+H]^+$ to $[M+Na]^+$ might be affected by the sodium concentration of a particular sample, which will make quantification difficult. In some cases, analyte ions might even lose their charge as caused by the reaction with a ligand present in the matrix which is capable in forming binary or tertiary complexes [2].

Signal suppression or enhancement is most efficiently reduced by proper clean-up procedures prior to injection into the LC-MS or by chromatographic separation of the analyte from the suppressing matrix compounds. An alternative interface might be tested if these approaches fail. Internal standards, unless they are isotopically labeled, are of limited value, since it is very difficult to find an analog which is equally affected by signal suppression as the analyte. In many cases, signal suppression or enhancement can be reduced but not eliminated. Such situations require the use of proper calibration techniques like the production of calibration solutions in blank matrix extracts instead of pure solvent. Such techniques are not free of potential pitfalls. In the case of multiresidue pesticide methods, it might not be easy to obtain negative samples. In some cases, matrix effects can even vary within one single matrix (e.g., veterinary drugs in urine) and might require the spiking of every individual sample in order to obtain unbiased results.

Linear Dynamic Range   LC-MS (TOF and quadrupole) calibration functions can be perfectly linear, while some others might be better dealt with a quadratic calibration function. Deviation from linearity is not a problem with modern calibration software, as long as non-linearity is not associated with a corresponding mass shift. Problems can occur if calibration curve shapes in neat standard solutions are different from those in spiked matrix. This has been observed for analytes producing sodium adducts. Depending on the availability of sodium in the sample, the ratio $[M+H]^+/[M+Na]^+$ will change.

Early TOF instrumentation was plagued by a narrow dynamic range of the detector. This has caused the reputation among some scientists, that TOF is not a quantitative technique. Although present TOF technology does not meet the performance of quadrupole instruments in this respect, TOF linear dynamic ranges of up to 3–4 magnitudes have been reported [3]. The linearity of TOF measurements is limited because of the way ions are detected. The accurate recording of the ion flight time is very challenging from the engineering point of view. The relatively short flight paths, as provided by bench-top instruments, require very accurate, high speed measurements. The recording chain, starting from the detector to the amplifier and the data storage has to work extremely fast. Currently, only micro channel plate (MCP) detectors are able to fulfill this requirement. Processing of the signal is the responsibility of an analog to digital converter (ADC) or a time to digital converter (TDC). Such converters are basically ion counting devices, which need a certain time period to "recover" from a striking ion, before the next arriving ion can be recorded. A real spectrum is produced by summing many hundred consecutive digital spectra. This is not an issue, since the time required for an individual TOF scan is incredibly short, permitting the addition of many consecutive spectra, while still providing excellent resolution in time. Many TOF instruments use an ADC which permits the recording of multiple hits, e.g., 2^n, within a given time bin. This improves the dynamic range up to 4 orders of magnitude.

The described limitations produced by the dynamic range could simply be solved by nonlinear calibration functions. However, the real problem is associated with mass shifts which are observed, when reaching detector saturation. Even a swarm of identical ions will hit the detector with a minimal band spread as caused by some variation of initial transition energy and imperfections during acceleration. A quantitative and a qualitative problem results if two ions arrive within a shorter time interval than the detector can "recover" from the first hitting ion. The detector will only record the time of the first ion. The second ion is neither counted nor is the flight time known. Consequently the average flight time is biased, since only the first hitting ions are available for calculating average flight times and finally exact masses. Hence overloading of a TDC detector causes mass shifts toward lighter masses. Modern TDC-based TOF instruments employ a number of correction algorithms which have reduced but not completely solved this problem. TDC detectors, which are much more prone to overloading than ADC detectors, are still being used. An approach to counter their limitation was implemented in instruments manufactured by Waters. The employed technique, called dynamic range enhancement (DRE), alternately records a normal ion beam as well as an attenuated ion beam. The attenuation is achieved by changing the potential of an ion lens within the interface. The increase of this voltage causes a defocusing of the ion beam and consequently reduces the number of ions passing the aperture and reaching the pusher region. This lens potential is varied in short intervals, producing a sequence of pulses of unattenuated and attenuated beam sections. The employed software algorithm utilizes unattenuated signals and "stitches in" an attenuated signal, if the ion count of one particular mass reaches saturation. The signal intensity of the stitched in attenuated mass peak is beforehand multiplied by the known magnification factor (ratio unattenuated/attenuated signal). This permits the calculation of spectra free of mass distortion over a significantly larger dynamic range. If carefully calibrated, linear dynamic ranges of 3–4 magnitudes can be obtained, which are similar to the reported performance of ADC instruments. Such an approach might sound more cumbersome than the more straightforward way of ion detection as provided by an ADC. However, the described limitation of a TDC is countered by the significantly higher data recording speed as compared to a slower ADC. This is not very relevant for ions beyond a mass range of 500–800. However, lighter (faster) ions are recorded by a TDC with a higher mass resolution than by an ADC. Ions with a mass of 200 might be only resolved at approximately half the resolution of heavy ions, if an ADC is employed [4]. On the other hand, resolution of a TDC based instrument is nearly constant over the relevant mass range. This can be a relevant issue if light molecules like pesticides and some veterinary drugs are to be monitored in difficult matrices. Light ions are not only faster in the field free flight tube but also in the ion optic of the interface. Given the finite dimension of the pusher and the pulse nature of oa-TOF, light ions are detected with a poorer duty cycle than heavier ions. These problems multiply if light polar ions elute at low concentrations, close to the dead volume from a chromatographic column where many matrix compounds appear. Negative effects regarding selectivity and sensitivity are consequently to be expected.

The issue of ADC versus TDC is a highly controversial issue within the TOF community. The future direction is not yet clear and will be determined by emerging engineering progress. It is even possible that instruments with a combination of both techniques will become available in the future.

Stability of Exact Mass Measurements   Maintaining the stability of the mass axis is most important for any high resolution and exact mass MS. Mass deviations of 2–5 ppm are achievable nowadays with modern bench-top TOF instruments. However, exact masses tend to drift, as caused, for example, by minor expansion or contraction of the flight tube, induced by temperature changes in the laboratory. Depending on the stability of the instrument, a recalibration before injecting a sample might be sufficient. Other instruments require the use of a continuous internal lock mass or the periodic recalibration by switching a discontinuous lock spray. Calibration immediately before an injection is clearly the most convenient approach, if the used instrument design can reduce mass drifts during the length of the chromatographic run. Lock masses can efficiently compensate long- and short-term mass drifts. Both log mass approaches (continuous and discontinuous) have advantages and disadvantages. These techniques utilize a secondary sprayer which infuses the lock mass solution into the interface region. In the case of the continuous mode, the lock mass is present in any spectra throughout the whole chromatogram and should therefore enable the mass axis recalibration for every scan.

However, there is a likelihood that analyte and/or lock mass ions cause signal suppressions. Furthermore, a chromatographic peak with an almost identical mass as the lockspray mass might elute somewhere in the chromatogram. Such a compound will not be properly quantified and will likely cause a mass shift. These aspects become increasingly relevant if difficult matrices with many endogenous compounds, like kidney or honey, are to be analyzed. Practical experience with lock spray solutions also shows that it is very difficult to infuse contamination-free calibrant solutions. Hence a chromatogram might contain more mass peaks than just the intended lock masses. Alternatively, a discontinuous lock spray can be employed. One commercial interface uses a baffle which periodically switches between two positions. Depending on the position, eluent from the column or the lock mass spray reaches the cone of the mass interface. This approach eliminates the potential risk of interferences by isobaric sample compounds, as well as the likelihood of signal suppression. However, the switching takes a certain time. Highly resolved chromatographic peaks as provided by ultra performance liquid chromatography (UPLC) can be partially distorted if their apex coincides with a lock spray switching. This can negatively affect the integration accuracy of such chromatographic peaks.

Centroiding and Resolution-Related Problems   High resolution TOF data can generate an enormous volume of data which has to be stored and processed. This is the main reason why TOF data are preferably stored in the form of centroids (in the form of a "stick") and not in continuum data points. Reconstructed ion chromatograms can be obtained from centroid or continuous data. Such mass traces can be extremely selective if narrow mass windows are utilized, even if they are extracted from a very busy total ion chromatogram. Furthermore, the signal to noise ratio of a target analyte peak significantly improves and the interfering peaks will disappear when narrower and narrower extraction mass windows are used. It is tempting to narrow down the extraction mass window because it hardly affects the absolute peak area of the target peak until a point is reached where splitting of the chromatographic peak will be observed. However, there is another side of this coin: The resolution of TOF is not as excellent as it might appear to the user, when playing with the mass extraction window.

The continuum mass signal with a finite signal mass peak width is represented by a centroid which has only a height but no width. This permits the selective extraction of analyte "a," which possesses almost the same exact mass as peak "b" from a chromatogram. If the correct exact mass of analyte "a" is extracted by using a mass window width smaller than twice the mass difference between "a" and "b," only analyte "a" shows up in the reconstructed ion chromatogram. Vice versa, only analyte "b" appears if the exact mass of this analyte with a mass window less than twice the mass difference between analyte "a" and "b" is extracted. However, a problem arises if these two analytes happen to co-elute in the chromatogram. The actual TOF resolution might not be as high as to physically resolve these peaks and the centroiding algorithm will likely represent the two analyte masses by a single centroid whose mass is a weighted average of the individual exact masses of analyte "a" and "b."

Such a potential interference was simulated by flow injecting a solution containing two compounds of almost equal exact mass [5]. Varied was the relative concentration of the two compounds to each other. If the two compounds have a mass difference larger than the mass resolution provided by the instrument, assigning centroids will be correctly performed. However, partial merged mass peaks, exhibiting significantly different relative heights, are often not well represented by centroids. Depending on parameters used to control the transformation into centroids, a wrong signal intensity, or worse, wrong exact mass assignments were observed. If a minor mass peak next to a large peak is to be extracted by using a narrow mass window, no signal might be obtained, since it was integrated "swallowed" into the centroid of its dominating neighbor. This data refer to a simulated condition where a complete co-elution was provoked (infusion experiment). The situation in a chromatogram is likely to be different. Such interference will be rare if high chromatographic resolution is used. However, they are still possible if trace analytes are to be determined in complex matrices containing many different endogenous compounds in large concentration. In such case, higher selectivity as provided by better resolving high resolution MS or triple quadrupole might be needed. This potential problem will remain an issue for LC-TOF until instruments of resolutions of some 30,000 FWHM will become available. Using current technology, residue analysis should be performed with the best available chromatographic resolution, e.g., UPLC, rapid resolution.

Another approach to prove the absence of such potential interfering compounds is the spiking of the sample with a small concentration of analyte. A significant deviation from the expected recovery would be expected if such interferences are present.

Selectivity Issues As indicated by the number of published works, MS-MS is clearly preferred over unit resolving single stage quadrupole for residue analysis, involving difficult food matrices. Figure 9.1 compares the selectivity obtained by a single stage and triple stage quadrupole when analyzing a fish sample spiked with a relevant level of oxytetracycline. This clearly indicates that MS-MS provides an enormous selectivity and consequently sensitivity gain. Figure 9.2 shows chromatograms of the same sample as in Figure 9.1. Measurements were done by LC-TOF. The chromatogram at the top in Figure 9.2 monitors $[M+H]^+$ based on a wide mass window of 1 Dalton. This corresponds to the resolution provided by a quadrupole

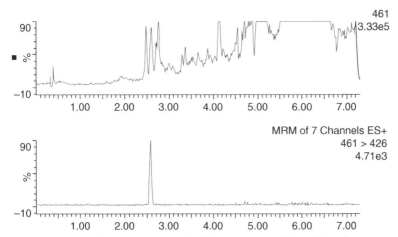

Figure 9.1.   UPLC Chromatogram of a fish sample containing 100 µg/kg oxytetracycline. The detector utilized was a triple quadrupole. The chromatogram at the top shows the trace of the unfragmented $[M+H]^+$ ion (the second quadrupole was set to rf only).
The chromatogram at the bottom shows an MS-MS (MRM) transition. Utilizing MS-MS improves selectivity and sensitivity as indicated by the presence of an oxytetracycline peak free of any interference.

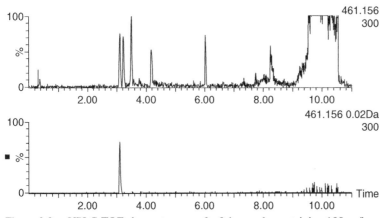

Figure 9.2.   UPLC-TOF chromatogram of a fish sample containing 100 µg/kg oxytetracycline. The chromatogram at the top uses a mass extraction width of 1 Dalton, resembling the performance of a unit resolving quadrupole. The chromatogram at the bottom utilizes a mass extraction width of only 0.02 Dalton. Clearly visible is the significant improvement in selectivity and sensitivity when reducing the width of the mass window.

MS. Hence, this chromatogram resembles the single stage quadrupole chromatogram as shown in Figure 9.1. The chromatogram shown at the bottom of Figure 9.2 was produced by extracting the same ion based on a narrow mass window of 0.02 Dalton. Clearly visible is the improvement of selectivity as compared to the trace depicted above.

9.1.2 Requirements for Quantitative Veterinary Drug Residue Methods

Concentration Range of Target Analytes   A significant challenge faced by veterinary drug residue analysts is the low reporting level of some banned veterinary drugs. Drugs like nitrofurans, chloramphenicol and nitroimidazoles have to be quantified and confirmed at concentrations at or below 1 μg/kg. Many permitted drugs have MRL concentrations of 100 μg/kg. Depending on the organs analyzed, levels up to 2000 μg/kg have to be quantified. The coverage of such a large dynamic range requires the latest generation of TOF instruments.

Clean-Up and Interfering Endogenous Compounds   The major difficulty related to the analysis of veterinary drug residues is caused by endogenous compounds which are present in the matrix at much higher concentrations than the target analytes.

The analysis of veterinary drugs in meat matrices is probably more difficult than the analysis of pesticides in vegetables or fruits because of the presence of high protein and fat concentrations.

Multiresidue clean-up protocols represent always a compromise between acceptable analyte loss and sufficient clean extracts. The more drugs (showing different polarity and pK values) to be covered, the more limited the degrees of freedom for clean-up procedures. Hence clean-up has to focus on the removal of proteins, fat, and carbohydrates from the matrix. A variety of extraction protocols are currently needed to analyze all the veterinary drugs which are in use. Some compounds are highly polar (e.g., aminoglycosides) while others are non-polar (e.g., avermectins, ionophores). Acceptable recovery of an incurred drug will only be achieved if the polarity of the extraction solvent corresponds to that of the analyte. Hence extraction with a single solvent will limit the range of recoverable analytes. Furthermore, many veterinary drugs are bound to the matrix by weak or strong interactions. Aminoglycosides can only be quantitatively liberated by low pH conditions, while nitrofuran metabolites are even covalently bound to the matrix. Unfortunately, a low pH extraction medium, as required for liberating aminoglycosides and nitrofurans, will decompose many pH labile compounds like penicillin's. This is the reason why no unified extraction protocol has yet been reported to be capable of recovering all relevant classes of veterinary drugs.

The high protein content of meat matrices is responsible for a number of problems. Proteins can precipitate during solvent exchange, or even in the HPLC vial, which can lead to the loss of analytes and the fouling of the analytical column. Some cell compartmented enzymes liberated during homogenization can cause significant analyte degradation if no suitable measures are imposed. Other proteins or phospholipids are responsible for interface-related signal suppression effects.

Their required removal from the extract is challenging because of the heterogeneous nature of proteins. Solid phase extraction (SPE) is an efficient way to remove parts of interfering proteins and to concentrate the desired analytes in the elution fraction. While there are very successful clean-up procedures for some veterinary drugs by cation or anion exchange SPE, such a treatment is too specific for a multiresidue method which should cover basic, neutral, and acid analytes. More suited are generic reversed phase SPE phase which—under optimized conditions—retain most drugs with the exception of very polar analytes like aminoglycosides.

Validation Procedures   There is no fundamental reason to apply a significantly different validation protocol to an LC-TOF method than to a LC-MS-MS or LC-Fluorescence method. However, the above stated limitations of TOF have to be critically evaluated during validation. This includes tests concerning the selectivity. The EU validation guideline (Commission Decision 2002/657/EC) demands the testing of negative samples reflecting the natural variety within a given matrix. In the case of animal tissues, this should include tissues from female and male, young and old animals. Other aspects to be tested are feeding regimes (e.g., lowland or alpine environment, conventional or organic feeding). Obviously, the organization of such samples that are free of any residues can present a significant logistic problem. Blank samples producing a relevant peak area for a given analyte raises the question of whether the observed signal is related to a false positive signal or an analyte trace. More often than not, no sufficient sensitive confirmatory method is available for providing an answer.

Many published LC-MS-MS methods were validated according to the commission decision; however, few LC-TOF reports are available [6]. Analytical performance criteria like selectivity, sensitivity, and linear dynamic range of a LC-TOF method will be strongly dependent on the chosen mass extraction window. The proper setting of this parameter has to be carefully evaluated and clearly specified in the method. Currently there is no agreement about a "proper" value of the mass extraction window. This will certainly change when more quantitative TOF methods are published. The mass extraction window is certainly strongly interrelated with the mass resolution of the instrument. However, the stability of the mass axis, required dynamic detection range, and the difficulty of the investigated matrix have to be considered as well. Testing for potential isobaric interferences is a prudent precaution. This can be done by a low and a high level spike into the injection-ready blank sample. A strong relative response deviation between these two solutions might indicate potential problems.

A TOF basically counts ions, hence the detector shows not an analog but rather a digital noise characteristic. Narrowing the mass extraction window will often lead to a situation where the recorded peak seems to grow directly out of a noise free baseline. The calculation of a meaningful value for sensitivity by signal to noise measurement is not possible under such circumstances. The European Union (Commission Decision 2002/657/EEC) proposed other sensitivity criteria, like the detection capability and decision limit ($CC\alpha$ and $CC\beta$), which describe the concentrations of an analyte above which a false positive, and false negative, respectively, finding can be ruled out. Depending on the calculation used, meaningless values of zero will result. Certainly TOF shows no indefinite sensitivity. Lowering the injected amount of analyte will decrease the peak size while the relative standard deviation of the measured peak area grows dramatically. We suggest determining $CC\beta$ by repeated injection of a series of analyte solutions with decreasing concentration. $CC\beta$ is reached if the calculated sum consisting of the signal produced by a blank plus

2.33 times the standard deviation of the investigated spiked sample reaches the signal intensity of the spiked sample.

9.2 METHOD

9.2.1 Instrumental

Chromatographic separation was achieved by using an UPLC system (pump and autosampler; Waters, Milford, MA), and a T3 HSS; 1.8 μm, 2.1 · 100 mm column (Waters). The mobile phase A consisted of 5% acetonitrile and 0.3% formic acid in water. Mobile phase B contained 5% water and 0.3% formic acid in acetonitrile. A linear gradient was applied: 0 min: 0% B; 2 min: 0% B; 8 min: 30% B; 12 min: 100% B; 13 min: 100% B; 13.01 min: 0% B; 14 min: 0% B; 14.6 min: 0% B. The flow rate was set to 0.4 mL/min and the column temperature to 30 °C. Injection volume was 6 μL.

The mass spectrometer used was an LCT Premier from Waters. The ESI interface was operated in the positive mode and the capillary maintained at 3 kV; cone voltage: 50 V; source temp: 150 °C; and desolvation temp: 350 °C. Gas flow was set to 640 L/h. Cumulated spectra (m/z: 100–1000) were taken every 0.2 seconds. The mass axis was corrected by a lock spray (leucine enkephaline) which was activated after every 30 scans. The instrument was operated in the dynamic range enhanced mode (DRE). Mass windows of 60 ppm were used to extract mass traces.

9.2.2 Analytes

Chinolones: ciprofloxacin, danofloxacin, enrofloxacin, flumequin, oxolinic acid, enoxacin, lomefloxacin, nalidixic acid, norfloxacin, ofloxacin, difloxacin, sarafloxacin, piromidic acid, sparfloxacin, azithromycin

Sulfonamides: sulfadiazine, sulfathiazole, sulfapyridine, sulfamerazine, sulfamethizole, sulfadimidine, sulfamethoxypyridazine, sulfachlorpyridazine, sulfachlorpyrazine, sulfadoxine, sulfadimethoxine, sulfacetamide, sulfamethoxazole, sulfisoxazole, sulfabenzamide, sulfameter, sulfamonomethoxine, sulfamoxole, sulfanitran, sulfaquinoxaline, sulfasalazine, sulfisomidine

Tetracyclines: Oxy-tetracycline; tetracycline; chlor-tetracycline; minocycline; doxycycline; Demeclocycline

Nitroimidazoles: Ipronidazole-OH, metronidazole-OH, dimetrinidazole, ronidazole, metronidazole, ipronidazole, HMMNI, tinidazole

Cephalosporines: Cefoperazone, cefazoline, cephalexin, cephapirin

Macrolides: Roxithromycin, tylosin A, erythromycin A, tilmicosin, spiramycin I, II, III, oleandomycin, josamycin, tulathromycin

Lincosamides: Lincosamide, lincomycin, pirlimycin, iso-pirlimycin, clindamycin, tiamulin

Benzimidazoles: Febantel, albendazole, fenbendazole, fleroxacin, flubenda-zole, mebendazole, ofendazole, oxibendazole, parabendazole, thiabenda-zole, triclobendazole

Penicillins: Ampicillin, nafcillin, penicillin G, penicillin V, dicloxacillin, cloxacillin, amoxicillin, oxacillin

Tranquilizers: Acepromazine, azaperol, azaperone, carazolol, chlorpro-mazine, propionylpromazine, xylazine

Various: Acriflavine, eprinomectin, praziquantel, trimethoprim, diaveridine, rifampicin, pyrimethamin, rifamixin, virginiamycin

9.2.3 Sample Processing

A portion of 6 g of meat sample is homogenized with 30 mL of acetonitrile. After 5 minutes 1 g of ammonium sulphate and 30 mL of extraction solvent (5.9 g succinic acid dissolved in 1 L water and adjusted to pH 5 by ammonium hydroxide) is added before homogenization continues. The solution is centrifuged (10 minutes at 8000 rotation/minute). The supernatant to which additional 2 g of ammonium sulphate is added is evaporated under vacuum by using a syncore device (Büchi; Flawil, Swit-zerland). The remaining aqueous solution is adjusted to pH 6.5 by adding ammonium hydroxide and afterwards centrifuged (5 minutes, 14,500 rotation/minute). The emptied evaporation vessel is rinsed with 6 mL of 50% aqueous dimethylsulfoxide (DMSO). This rinsing solution is centrifuged (5 minutes at 14,500 rotation/minute). A 200 mg Oasis HLB SPE cartridge (Waters) cartridge is activated by 3 mL ace-tonitrile and 2 mL water. The centrifuged rinsing solution is passed through the preconditioned cartridge, followed by 5 mL of water. Afterwards, the evaporated, pH adjusted, and centrifuged extraction solution is given onto the SPE cartridge. Elution of the SPE cartridge into a pre-weighted conical vial is done by 2 mL ace-tonitrile and 3 mL succinate extraction solvent / acetonitrile (1/2). Then, 0.4 mL DMSO is added into the vial and the solvent mix is evaporated in a heating block 40 °C by a stream of air. The evaporation process is stopped when the remaining liquid is about 0.5 mL. Water is added until the net weight in the vial reaches 2 g. The obtained, clear sample is ready for injection after centrifugation (5 minutes at 4,000 rotation/minute). A typical chromatogram (total ion current) is shown in Figure 9.3.

9.2.4 Validation Data

Validation was done by five different spiking levels. Depending on the MRL of the analyte, 1; 3.16; 10; 31.6; 100, respectively 10; 31.6; 100; 316; 1000 μg/kg spikes were added. Each spiking level was repeated four times. Such series were performed three times on different days by different operators. The EU guideline was followed by additionally extracting and analyzing 20 different blank samples to test for pos-sible interfering compounds.

The whole protocol was performed for three matrices: meat, kidney and liver.

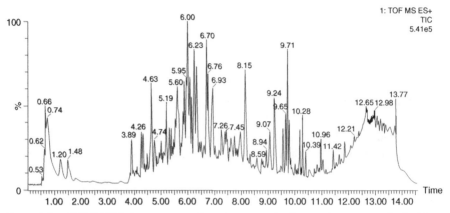

Figure 9.3.   UPLC-TOF chromatogram (total ion current) of a muscle sample spiked with veterinary drugs and processed according to the described method.

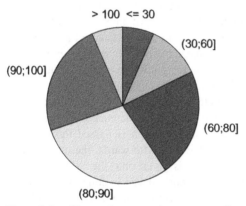

Figure 9.4.   Observed recovery in percentage for all veterinary drugs covered by the proposed method. The slice size is proportional to the number of analytes recovered within a given range as indicated by the adjacent numbers.

Analyte recovery was rather high; more than 60% of all compounds were recovered with rates of at least 80% (see Figure 9.4). High recovery rates were obtained in all tested matrices, indicating that the sample extraction and clean-up is well suited for a multiresidue method. The coefficient of determination r^2 related to the five spiking levels in muscle was larger than 0.99 for 50% of all compounds (see Figure 9.5). There is a clear relationship between recovery and r^2 which indicates that poor r^2 values were mostly linked to poorly extracted compounds. Another reason for lower r^2 is the fact that the apex of some analyte peaks fell within the timing of the lock spray switching. The relatively slow switching of the baffle and less the actual lock mass measuring time was found responsible for this limitation. This observation refers to the older LCT Premier instrument which was used for this study.

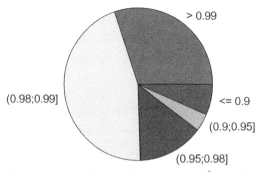

Figure 9.5.   Coefficient of determination r², describing the spiking of muscle samples by five different concentration levels (two orders of magnitude) and four repetitions per level. The slice size is proportional to the number of analytes producing coefficient of determination as indicated by the adjacent numbers.

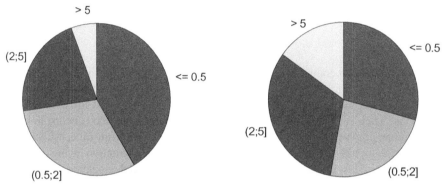

Figure 9.6.   Sensitivity (detection capability CCα) of the various veterinary drugs in muscle and liver. The slice size is proportional to the number of analytes being detected within he specified sensitivity range (µg/kg).

The obtained sensitivity is a product of a high enrichment factor during sample processing and the narrow UPLC peaks. Forty percent all analytes in muscle can be detected at ≤0.5 µg/kg (detection capability CCα). Only 6% show poorer sensitivities than 5 µg/kg. The performance in liver and kidney is slightly poorer concerning this aspect (see Figure 9.6). The reason for this is explained by potential isobaric inter-ferences. Some chinolones show CCα of 20 µg/kg in liver, while 10–20 times lower values are typical for the muscle matrix. Most other analytes are detectable at similar concentration in muscle, liver, and kidney. This limitation is not very critical since MRL levels of most compounds are significantly higher for liver and kidney than for muscle. Signal suppression is an issue and requires the use of matrix-spiked calibration solutions. Only 3% of all compounds (muscle matrix) and 6% of all compounds (liver matrix) showed extensive signal suppression. Extensive signal suppression refers to analytes which produce less than 25% signal intensity in matrix as compared to pure standard solution) (see Figure 9.7).

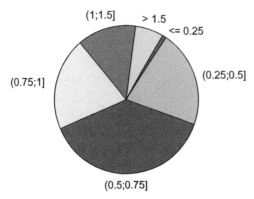

Figure 9.7. Signal suppression in muscle tissue observed for the various analytes. The slice size is proportional to the number of analytes showing suppression within the specified sensitivity range as indicated by the adjacent numbers. Suppression is defined as ratio of the peak area of a fixed analyte concentration in matrix and the peak area of the same concentration in pure solvent.

A total of 106 compounds were considered to be quantified by the proposed method. Originally some more compounds were included but had to be removed from the list because of various problems. Most avermectines and ionophores are too apolar. Their recovery from the matrix is low; furthermore, their inclusion into the chromatographic method would significantly prolong the run time. Malachite green is a quaternary amine which prevents the reproducible retention by the utilized reversed phase SPE. Erythromycin undergoes partial thermal degradation during the evaporation step. The degradation product (loss of water) is detectable; however, because of a lack of reference substance, quantification is difficult. A few compounds seem to show partial enzymatic degradation in liver and kidney, resulting in poorer quantification performance. The detection limit for some of the banned nitroimidazole drugs is close to 1 μg/kg which is critical since these compounds should be detected at that level. The performance for these compounds can be improved by avoiding DMSO as keeper (isobaric interferences) and employing a steeper initial gradient to sharpen the peaks of the early eluting nitromidazoles. However, avoiding DMSO will reduce recoveries of some polar analytes.

9.3 DISCUSSION

9.3.1 Extraction and Clean-Up of the Sample

The proposed multi-residue method extends the range of quantifiable drugs to some 100 different analytes, yet it does not include aminoglycosides and nitrofurans. Recoveries of analytes were significantly improved by an approach we termed bi-polarity extraction. This concept uses a mixture of acetonitrile and an acidic aqueous buffer. The intermixing of these two solvents is prevented by a high concentration of ammonium sulphate. Hence the homogenization of the sample is done with an extraction solvent emulsion showing a significant polarity difference. The extract is centrifuged and the two liquid layers are transferred into an automated vacuum evaporation device in order to remove all traces of organic solvent. This solvent exchange step induces protein precipitation which is responsible for partial losses of apolar analytes. Hence the evaporation vessel and the precipitates are rinsed by an organic solvent, which after dilution can be transferred into the SPE cartridge. The main extract, containing the polar, weakly retained analytes, is added afterwards. After the elution from the SPE cartridge, another solvent exchange step is required. Evaporation in a heating block under a steady flow of air until dryness was associated again with irreproducible loss of some analytes. Hence evaporation was stopped by the use of a keeper. Dimethylsulfate (DMSO) has a strong solvation power and a high boiling point, which makes it an ideal keeper. Even more important is the fact that DMSO shows a relatively weak eluting strength. While even a low concentration of residual acetonitrile in the final extract ruins the chromatography of early eluting peaks, much higher concentrations of DMSO can be tolerated in the injection solution. This technique significantly reduces losses during the final solvent exchange step. The evaporated extract is diluted with water and centrifuged to produce a precipitation-free solution. This ensures proper UPLC separation and prolonged column life.

9.3.2 High Chromatographic Separation Efficiency

TOF has profited more than probably any other MS technique from the introduction of sub 2-μm particulate HPLC column packing materials. Sub 2-μm stationary phase packed in columns with 10 or more centimeter length permit extremely powerful chromatographic separations, resembling GC separations as produced by fused silica columns. Advanced instrumentation like ultra performance liquid chromatography (UPLC) is capable of producing the required pressure for delivering the mobile phase. Peak width is reduced, while the peak area is maintained, resulting in an absolute peak height increase. This can produce sensitivity gains of a factor of 2–4 [7]. The limited selectivity of TOF (as compared to MS-MS) can be significantly improved by increased chromatographic resolution. UPLC is a powerful tool, yet it requires a fast detector to follow the narrow chromatographic peaks. Triple quadrupole instruments rely on defined dwell times to monitor analytes. This is not a limitation if only few analytes are to be monitored, however, it becomes a potential problem for multiresidue methods. Although modern MS-MS instruments have been significantly improved, the shortening of dwell times negatively affects the sensitivity and reproducibility. TOF being inherently faster than MS-MS can detect a theoretically unlimited number of analytes without compromising sensitivity and selectivity. Hence, the coupling of UPLC with TOF is highly beneficial for multiresidue methods where difficult matrices are involved. The higher chromatographic resolution narrows not only analyte peaks but also suppression windows, hence reducing the likelihood of signal suppression. However, often the responsible agents for suppression are proteins or phospholipids which can elute as a broad hump.

9.3.3 Detection and Data Processing

The processing of TOF data still presents a major challenge. TOF chromatograms require much more computer storage capacity and calculation power than triple quadrupole data. Before buying an instrument which is supposed to analyze several hundred compounds in one chromatographic run, a suitability evaluation of the supplied software is suggested. Critical points are processing speed and user friendliness. Analyzing 100 analytes in a complex matrix will often include the verification and confirmation of a peak which might be a false positive signal. This requires a look at the underlying spectra. Does it happen that this chromatographic peak is a false positive signal, because it is produced by an isotope of a coeluting endogenous substance? Can it be that there is the likelihood of a coeluting isobaric compound? Can we confirm the presence of the peak by locating its typical sodium adduct or possibly by a metabolite of the parent drug? The evaluation of chromatograms can become very slow if the utilized software is not providing tools designed for providing quick answers to such questions. Hence the processing of data currently presents a major bottleneck. It is to be expected that faster computing power and hard-disk access as well as new dedicated software will help the user to perform this task in a shorter time and with greater confidence.

Confirmation of positive findings can be partially done by checking the presence of adducts and metabolites. However, it is still advisable to use a triple quadrupole for confirmation of such samples. This is not much of a limitation, since the same sample extract can be injected. Unlike in the field of pesticide analysis, the number of positive findings for veterinary drugs in animal-based samples is rather low. Hence the number of MS-MS confirmations is manageable.

9.4 CONCLUSIONS

The use of LC-TOF for multiresidue methods in the field of veterinary drug residues is still very recent. As far as we are aware, the proposed method is the first published method where an extensive array of analytes was validated according to the EU guidelines. Although other groups are working in the same field, practical experience is still limited.

It is always difficult to predict future developments and trends. More often than not, some unexpected developments and technological advances will alter envisioned trends. However, triple quadrupole is a technology which has reached a certain level of majority. Progress (shorter dwell times, higher sensitivity and possibly higher resolution) is still to be expected, although these gains will be less dramatic than in the past. MS-MS is probably close to the limits defined by the law of physics. On the other hand, TOF is still a niche product which has just become feasible because of advances occurring in other disciplines (e.g., fast electronics and computing power). TOF instruments became relevant for residue analysis after resolutions of some 10,000 FWHM were available. Maybe less visible for outsiders, there has been significant progress concerning the stability and user friendliness of TOF instruments in recent years. Currently, TOF is still far away from the boundaries defined by physics. There are successful concepts of how to improve the duty cycle, which will provide more sensitivity. Commercial TOF instruments are using 4 GHz digitizers while 10 GHz processing devices are available. Once utilized, this will improve the dynamic range and possibly resolution. Engineering and, to a lesser degree, physics are hindering the development of faster microchannel plate detectors (MCP). Hence more sensitivity, higher resolution, and more ruggedness are likely to be expected in the near future. This will make it likely that some applications which are now covered by triple quadrupole will be taken over by LC-TOF.

As a group familiar with LC-MS-MS and LC-TOF, we strongly appreciate two aspects offered by LC-TOF which most likely will not become available for LC-MS-MS instrumentation. Exact masses (the basis for the identification of an analyte) are well defined, they do not vary for a given analyte, regardless of the MS technology, or brand of instrument being used. This is not the case for MS-MS spectra. Although there has been intense research, there are still no interplatform MS-MS spectra libraries available. Unlike exact masses, MRM transitions (fragment masses and collision energies) cannot simply be copied from a publication without being adapted to the particular instrument available.

Last but not least, we appreciate the concept of "inject first; think later" which is offered by LC-TOF. After observing a triple quadrupole MRM signal, there are often additional questions which require set-up of additional experiments. Are other adducts available? Do we have other fragments? Are the metabolites of the detected parent drug present as well? Knowing that the detected drug is often co-administered with another drug to animals, the question might arise whether that second drug is present in the sample as well? We may become informed by other groups about a newly detected drug. This might make us curious about whether this drug has been already present in some of our samples which have been analyzed last month or last year. In many cases, such questions can only be answered by a technique which produces high resolution full scan data and permits a-posteriori data mining capabilities.

REFERENCES

1. Chamber, E.; Wagrowski-Diehl, D.; Lu, Z.; Mazzeo, J., Systematic and comprehensive strategy for reducing matrix effects in LC-MS-MS analysis, *J. Chromatogr. B*, **2007**, 852, 22–34.
2. Alvarez, E.; Brodbelt, J., Evaluation of metal complexation as an alternative to protonation for electrospray ionization of pharmaceutical compounds, *Am. Soc. Mass Spectrom.*, **1998**, 9, 463–472.
3. Ferrer, I.; Thurman, E.M.; Fernández-Alba, A., Quantification and accurate mass analysis of pesticides in vegetables by LC/TOF-MS, *Anal. Chem.*, **2005**, 77, 2818–2825.
4. Hidalgo, A.; Fjeldsted, J.; Frazer, W., The application of high speed oscilloscope analog-to-digital converters to time-of-flight spectrometry, Poster ASMS (**2007**).
5. Kaufmann, A.; Butcher, P., Strategies to avoid false negative findings in residue analysis using liquid chromatography coupled to time-of-flight mass spectrometry, *Rapid Commun. Mass Spectrom.*, **2006**, 20, 3566–3572.
6. Hernando, M.; Mezcua, M.; Suárez-Bercena, J.; Fernández-Alba, A., LC-TOF for simultaneous determination of chemotherpeutant residues in salmon, *Anal. Chimica Acta*, **2006**, 562, 176–184.
7. Kaufmann, A.; Butcher, P., Quantitative liquid chromatography/tandem mass spectrometry determination of chloramphenicol residues using sub 2-μm particulate high-performance liquid chromatography columns for sensitivity and speed, *Rapid Commun. Mass Spectrom.*, **2005**, 19, 3694–3700.

CHAPTER 10

INDUSTRIAL APPLICATIONS OF LIQUID CHROMATOGRAPHY TIME-OF-FLIGHT MASS SPECTROMETRY

Jeffrey R. Gilbert, Jesse L. Balcer, Scott A. Young,
Dan A. Markham, Dennis O. Duebelbeis, and Paul Lewer

Dow AgroSciences, Indianapolis, Indiana

LIQUID CHROMATOGRAPHY time-of-flight mass spectrometry (LC/TOFMS) can be applied to solving a wide range of problems in industrial research and development, ultimately contributing to the discovery and development of new products. The ability of LC/TOFMS to address these problems has increased significantly over the last 15 years due to improvements in both LC separations and TOFMS instrumentation, which have improved the speed, sensitivity, resolution, and dynamic range of LC/TOFMS analysis. This chapter will explore these technical improvements, and how they have enabled LC/TOFMS to impact four areas of industrial R&D: 1) searching for biologically active molecules from natural sources; 2) identification of trace level agrochemical environmental degradates; 3) identification and quantitation of agrochemical metabolites of toxicological interest, and 4) characterization of recombinant proteins, including generation of detailed information on the primary, secondary, tertiary, and quaternary structures as well as post-translational modifications. Examples of state-of-the-art research in each of these areas will be discussed from our own work as well as from the literature.

10.1 INTRODUCTION

Since the advent of electrospray ionization (ESI) and atmospheric pressure chemical ionization (APCI) in the early 1990s [1, 2], LC/MS has impacted a wide range of both academic and industrial applications. LC/MS is now routinely used in many

Liquid Chromatography Time-of-Flight Mass Spectrometry: Principles, Tools, and Applications
for Accurate Mass Analysis, Edited by Imma Ferrer and E. Michael Thurman
Copyright © 2009 John Wiley & Sons, Inc.

areas of industrial research and development including; the search for new agrochemicals from nature, the identification of trace level environmental and toxicological metabolites, and the detailed characterization of proteins produced by plants and microbes. Early LC/MS systems were primarily based on either single or triple quadrupole mass analyzers, and were limited to providing low resolution MS and MS/MS data. In the mid-1990s, the advent of LC-TOFMS and LC-quadrupole time-of-flight (LC-QqTOFMS) instruments provided the first medium resolution (~10,000 FWHM) instruments for LC/MS analyses [3–5]. The improved resolution and mass accuracy provided by these TOF-based systems allowed researchers for the first time to obtain accurate MS and MS/MS data on trace level components in complex mixtures. This dramatically changed the types of industrial problems which could be addressed using LC/MS, especially in applications requiring compound identification.

Since their commercial introduction in the mid-1990s, several advancements have been made in LC/MS instruments incorporating TOF analyzers. The introduction of temperature compensation circuitry, as well as the use of TOF chamber designs incorporating thermal isolation, has improved mass accuracy by reducing the effects of thermal expansion on TOF mass measurements. Whereas early LC-TOFMS systems required the introduction of an internal standard into the LC eluent to obtain accurate mass data, more recent designs incorporating multiplexed "lock spray" and "dual sprayer" type interfaces allow the introduction of internal standards separate from the sample stream [6, 7]. This improves MS mass accuracy while minimizing suppression effects. Early LC-TOFMS systems utilized time to digital (TDC)-based circuitry to perform high-speed digitization of signals generated from the TOF multichannel plate (MCP) detectors. This TDC detection proved effective, but suffered from dynamic range limitations. More recently, LC-TOFMS instruments have been developed which utilize analog-to-digital (ADC) based digitization, greatly improving their linear dynamic range [8, 9]. This has allowed the use of LC-TOFMS in a broad range of both quantitative as well as qualitative applications [10, 11]. Finally, improvements in high-voltage power supplies and associated circuitry have allowed LC-TOFMS instruments to operate in the pulsed positive/negative mode, enabling a broader range of chemistries to be detected in a single experiment.

This chapter will explore the role of LC/TOFMS in several areas of industrial research which have benefited from these developments in LC-TOFMS instrumentation. These areas include: 1) screening natural product extracts to find new agrochemicals, 2) identifying trace level metabolites generated in environmental systems, 3) identification and quantitation of metabolites of toxicological interest, and 4) detailed characterization of recombinantly expressed proteins. The combination of high resolution, high mass accuracy, and high full scan sensitivity which can be obtained using LC/TOFMS has made it a valuable tool in each of these areas of research.

10.2   NATURAL PRODUCTS DISCOVERY— DEREPLICATION AND IDENTIFICATION OF NEW ACTIVES

Prior to the advent of LC/MS, the primary role of mass spectrometry in natural product discovery was in the structural elucidation of compounds isolated via bio-assay-guided fractionation. With the advent of commercial LC/MS instruments in the early 1990s, natural product chemists gained the ability to obtain mass spectra directly from complex natural product mixtures. This led to the use of LC/MS both for the identification of active compounds from natural sources, and as a tool for prioritizing extracts. Using LC/MS to screen for known bioactive molecules, a process often called dereplication, rapidly became a key role for this technology in natural product discovery. Many modern LC/MS dereplication systems use the basic configuration introduced by Constant et al. [12], which is shown schematically in Figure 10.1A. In this approach, crude extracts are separated with reversed-phase LC, and the column eluent is directed to a UV-VIS diode-array detector. The eluent is then split between a fraction collector and the mass spectrometer. The majority of the flow is directed toward the fraction collector and fractionated into 96-well plates, while a small percentage is ionized and detected with positive and/or negative ESI. This approach takes advantage of the concentration-dependent response of electro-spray [13] to provide sensitive molecular weight information, while fractionating and retaining the majority of the sample for subsequent bioassay. It also efficiently

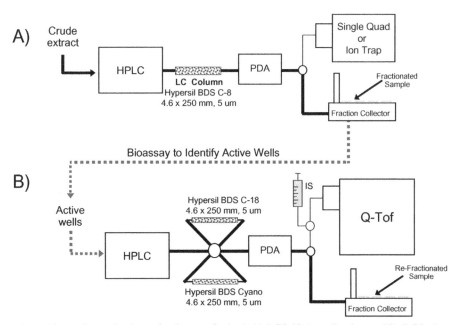

Figure 10.1.   General schematic diagram for both A) LC/MS dereplication and B) LC/MS identification systems.

provides corroborating structural data, since both ionization modes are often used in a single run, which can be crucial for trace-level samples. The resulting molecular weight and UV-Visible spectral data, combined with any known taxonomic information on the source, can then be used for database searching. If a database match is made, the extract is often de-prioritized, allowing the researcher to focus their activities on higher-value samples.

The introduction of commercial LC-TOFMS and LC-QqTOFMS instruments in the 1990s allowed natural product researchers to obtain both accurate MS and accurate MS/MS data directly from crude natural product mixtures. This provided a significant advancement in the dereplication and identification process, often allowing the unambiguous assessment of compound novelty using only LC/MS data. Where natural product database searches using nominal molecular weights often produced many possible matches (10–100), accurate mass-based searches generally produced a only a few (1–5) molecular formula matches [14, 15]. This was still true even when searching large databases such as the Chapman and Hall Dictionary of Natural Products (CHDNP), which currently contains over 210,000 entries of compounds isolated from nature [16].

Until recently, one limitation of LC-QqTOFMS instruments was their inability to perform pulsed positive/negative ionization on a scan-to-scan basis. As a result, a two-tiered LC/MS screening approach to dereplication and identification was often employed, as shown in Figure 10.1A-B [17]. In this system, initial dereplication of crude extracts was performed using a single quadrupole mass spectrometer system equipped with a UV diode-array detector (DAD) and fraction collector. This configuration allowed the acquisition of pulsed positive and negative electrospray mass spectra, both with and without non-selective source induced dissociation (SID). As a result, a single analysis could provide retention time, UV, positive ESI-MS, negative ESI-MS, as well as SID fragment ion data. Use of a pulsed positive/negative ionization-based approach has proven essential due to the large number of natural products which have been observed to ionize effectively in only one polarity mode via electrospray [17]. Data collected in the dereplication analysis were compared to an in-house spectral library of known bioactive compounds, and the collected eluent was bio-assayed to localize the region of bio-activity. A second injection was made on an LC-QqTOFMS system operated in data dependent MS/MS mode, with combined UV detection and fraction collection (Figure 10.1B). The accurate MS and MS/MS data from this system were used to search commercial databases using a combination of taxonomic, UV maxima, and accurate mass data. This approach often produced a single formula match, which could be confirmed by comparing the structural features of each library match to the observed accurate MS/MS fragments. If necessary, additional MS^n data could also be acquired by infusion of the collected 96-well plate contents into an ion trap configured with a nanospray source.

In one example, screening of a library of fungal extracts using a beet army worm (BAW, *Spodoptera littoralis*) assay highlighted strain MYCO-743-C5 as producing secondary metabolites with insecticidal activity. This extract was first analyzed on the LC/MS dereplication system described in Figure 10.1A, producing the chromatogram shown in Figure 10.2. Automated interrogation of the resulting

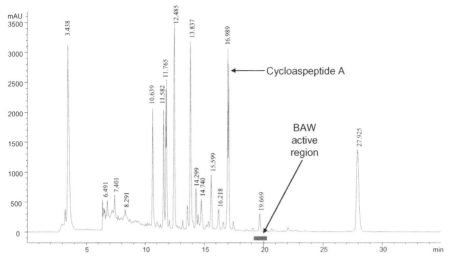

Figure 10.2.   LC chromatogram of insect-active fungal extract MYCO-743-C5.

UV and MS spectral data for this sample did not produce a match to any of the known actives in our internal spectral libraries. Therefore, BAW bioassay of the eluent from this run was used to localize the insecticidal activity to the 19–20 min region of the chromatogram.

The accurate MS and MS/MS spectra of the sample were then obtained via LC/MS on a Micromass Q-TOF configured as shown in Figure 10.1B. The bioactive region was found to contain two components, designated unknowns 1 (MW 633) and 2 (MW 625). The accurate mass of unknown 1 was measured on the Q-TOF as 633.2852 ± 0.005 Da. Searches of the CHDNP database for this mass produced two matches within ± 10 mDa, both with the formula $C_{37}H_{44}NO_6Cl$ (M = 633.2857). The UV and accurate MS/MS spectra for unknown 1 were consistent with one of these matches, penitrem A, which was a known insect-active. The structure of penitrem A, along with the measured UV, accurate MS, and accurate MS/MS for unknown 1 are summarized in Figure 10.3. These data were sufficient to allow the tentative assignment of unknown 1 as penitrem A, which was determined to be of no further interest.

The accurate mass of unknown 2 in extract MYCO-743-C5 was measured as 625.3291 ± 0.005 Da, as shown in Figure 10.4A. Searches of the CHDNP database produced no matches to this observed mass within ± 10 mDa, indicating that the unknown was a novel natural product. The MS/MS fragmentation pattern of the unknown, shown in Figure 10.4B, was consistent with a cyclic peptide structure. Further examination of the UV spectrum of the unknown indicated that it was related to cycloaspeptide A, which was subsequently detected in the extract (RT = 16.9 min) at a 60-fold higher concentration. Detailed examination of the accurate MS/MS spectrum of cycloaspeptide A was performed, and provided a framework for the assignment of all the major fragment ions of unknown 2 to deoxygenated cycloaspeptide A. The cyclic nature of the peptide allowed the observation of several

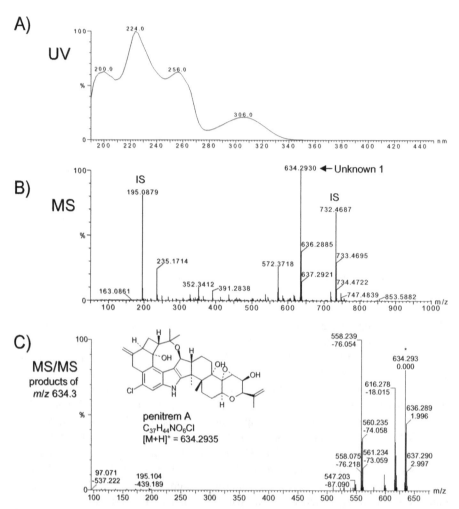

Figure 10.3. UV, MS, and MS/MS spectra of Unknown 1 (penitrem A).

fragmentation pathways, which are detailed in Figure 10.4C. The unknown was then chromatographically isolated, and its full structure was determined via NMR. The absolute stereochemistry of the constituent amino acids for the novel cycloaspeptide was also determined using the Marfey procedure [18] involving hydrolysis, derivatization with a chiral reagent, followed by LC/UV and LC/MS detection. In each case, the amino acids were found to belong to the L-series, confirming the structure to be consistent with the other known members from this family.

High throughput (HTS) screening of fungal and bacterial libraries typically occurs with limited taxonomic knowledge of the organism. Given the bioassay resources required to carry out bioassay-directed dereplication of crude extracts, along with the high potential for finding nuisance bioactive chemistry, interest has

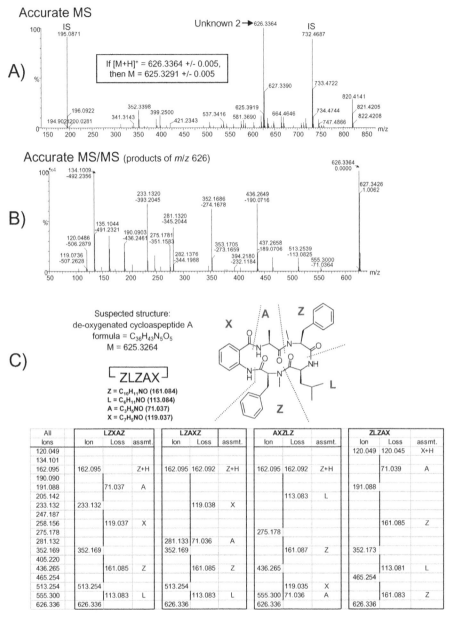

Figure 10.4.   Unknown 2, A) Accurate MS and B) MS/MS spectra, as well as C) proposed structure, and fragment ion assignments.

grown in developing methodologies to perform rapid non-bioassay-based evaluation of active extracts. One recent application of this approach utilized micro-scale LC-QqTOFMS to screen fungal extracts for mycotoxin-producing species [19]. This article provides useful positive ESI-MS and UV data for 474 mycotoxins and highlights the advantages of accurate mass for this type of application. In our experience, a combination of UV and accurate mass data is often useful to eliminate extracts unlikely to contribute toward the discovery of new lead chemistries. For example, a fungal extract was analyzed by LC/MS and produced multiple secondary metabolites with accurate masses of 556.1366 and 570.1520 Da. A search of CHDNP over the range of ±0.005 Da for these masses, combined with UV spectra, supported the rapid assignment of bi-naphthopyrandione as the active chemistry, which is produced by the genus *Aspergillus*.

10.3 ENVIRONMENTAL METABOLITE IDENTIFICATION USING LC/TOFMS

In order to register a new agrochemical, the degradation and metabolic fate of the compound in plants, animals, and the environment must first be determined. A variety of laboratory and field studies are performed to provide this information. These can include; plant and animal metabolism and nature-of-the-residue (NOR), aqueous photolysis, soil degradation (aerobic and anaerobic), hydrolysis, and water sediment studies. These studies typically generate complex samples containing trace level metabolites which must be identified. LC/MS plays a critical role in the structural elucidation and identification of these environmental metabolites. Samples generated in these studies frequently consist of complex matrices, making the detection of the resulting metabolites difficult. This identification is often accomplished using electrospray ionization combined with TOFMS to generate both accurate mass and MS/MS data. More recently, the combination of LC-TOFMS, nanospray ionization, and controlled isotopic labeling has been found to significantly speed the identification of these challenging environmental metabolites.

In one example, a new natural product-derived insect control agent, spinetoram, was under development due to its broad spectrum of insect activity, novel mode of action, and low persistence in the environment [20]. Spinetoram consists of two primary factors which are responsible for its insecticidal activity; the structure of the major factor is shown in Figure 10.5A. Spinetoram is readily ionized by ESI, and under MS/MS conditions produces fragment ions resulting from the loss of the sugar moieties (charge retention on the sugar fragment). These low mass fragment ions have proven difficult to observe using ion trap-based instruments, since commercial ion traps are generally unable to detect fragment ions at less than 20–30% of the precursor mass [21]. These MS/MS fragment ions provide key information for the identification of spinetoram metabolites, and can be readily observed using LC-QqTOFMS instruments.

To further aid in the identification of the trace level spinetoram metabolites present in complex environmental samples, a specialized spinetoram isotopic labeling scheme was developed. By combining both ^{14}C and deuterium-labeled forms of spinetoram, a unique isotopic "fingerprint" was produced which could be readily distinguished using mass spectrometry. Three forms of the major factor of spinetoram were combined in an approximate 1:1:1 (w/w/w) ratio mixture for use in these studies. While all three forms were uniformly labeled (^{14}C) throughout the macrolide portion of the molecule, two were also deuterated at either the ethoxy

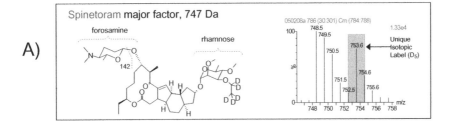

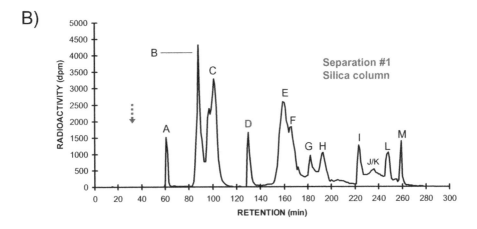

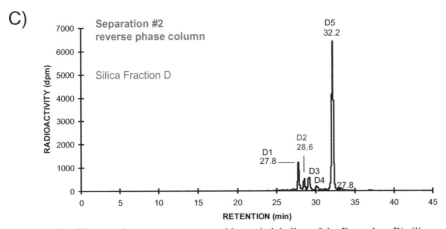

Figure 10.5.   The A) spinetoram structure and isotopic labeling of the D_5 analog, B) silica column radiochromatogram, and C) reverse phase separation of silica column fraction D generated in a lettuce NOR study.

group on the rhamnose sugar (D_5) or at the 5–6 positions on the macrolide ring (D_2). The resulting $[M + H]^+$ isotopic cluster of the labeled spinetoram major factor is shown in Figure 10.5A. Assuming that the isotopically labeled portion of the molecule remains intact, the mass spectra of all spinetoram metabolites generated in the subsequent environmental studies must contain this unusual isotopic cluster. In the subsequent experiments, this isotopic pattern was used to confirm that the observed masses were related to the applied material.

A probe study was conducted to gain an initial understanding of the metabolic fate of the major spinetoram factor in lettuce. This study was conducted using the labeled test material described above. Spinetoram was prepared as an emulsifiable concentrate (EC), and diluted with water to a concentration of 0.75 mg/mL. This was sprayed onto leaves at a rate of 900 g/ha (9 µg/cm^2), and allowed to metabolize for a period of 0.25–21 days. The 3-day samples were extracted with water/acetonitrile, and initially separated using a preparative silica (0.7 × 100 cm) gel column. A gradient elution from hexane to methanol was performed, and fractions were collected and assayed for ^{14}C. The resulting radio-chromatogram is shown in Figure 10.5B.

As this chromatogram illustrates, the lettuce experiment produced a complex set of metabolites, with more than 12 radiochemical peaks observed in the silica column separation. Fractions from one of these peaks, designated as peak D in Figure 10.5B, were then combined and analyzed on a separate gradient reversed-phase LC system. The eluent from the reversed-phase separation was fractionated into a 96-well microtitre plate, and subsequently analyzed for radiochemical content using a Packard TopCount NXT liquid scintillation counter. The chromatogram produced using this reversed-phase separation is shown in Figure 10.5C. Several radiochemical peaks (D1–D5) were observed in the 25–35 min region of reversed-phase chromatogram from silica fraction D. One of these, designated as peak D2 (RT = 28.6 minutes), was subsequently collected for nanospray. Fraction D2 was infused using a pulled-capillary nanospray source coupled to a Sciex QSTAR XL in an attempt to identify the metabolites present, as described below.

The positive nanospray full scan spectrum obtained from nanospray infusion of component D2 is shown in Figure 10.6A. The abundance of charge provided by the nanospray ionization process produced a complex full scan mass spectrum in the region from m/z 100 to 1000. Upon careful examination of the m/z 700–800 region, a peak at m/z 762.5 (i.e., 14 mass units greater than the parent spinetoram factor) was observed to contain an isotopic pattern matching the applied spinetoram material. This is shown in the inset of Figure 10.6A. MS/MS of m/z 762.5 was acquired in an attempt to further characterize this unknown, and the resulting MS/MS spectrum is shown in Figure 10.6B. Key ions observed in this MS/MS spectrum include m/z 203, and 156, corresponding to the intact trimethyl-rhamnose sugar, and modified forosamine sugar, respectively. The observation of a fragment at m/z 156, which is 14 mass units greater than the corresponding forosamine fragment ion from parent spinetoram, indicated that metabolism had occurred at this region of the molecule. One possible explanation for this shift in fragment ion mass was conversion to the N-formyl metabolite shown in the inset of Figure 10.6B. The N-formyl metabolite was subsequently synthesized, and found to produce a reversed-phase LC retention time match as well as both MS and MS/MS spectral matches to the

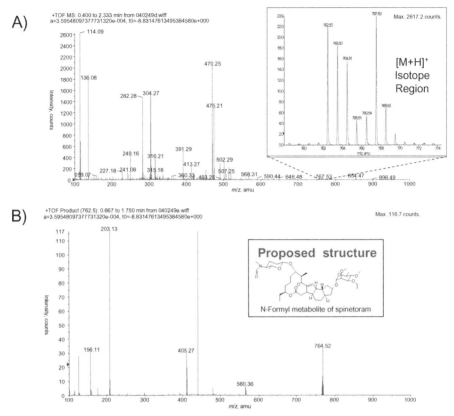

Figure 10.6. Nanospray A) MS and B) MS/MS spectra of a spinetoram metabolite (reversed phase fraction D2, retention time 28.6 min).

unknown peak. This confirmed the identification of this metabolite as the N-formyl metabolite of the applied material. Using this combination of conventional LC/MS, accurate MS and MS/MS, nanospray ionization, and controlled isotopic labeling, enabled the identification of more than twenty metabolites across three different nature of residue studies.

10.4   APPLICATIONS OF LC-TOFMS IN TOXICOLOGY

In additional to environmental studies, a complex series of animal metabolism studies are also required in order to register a new agrochemical. These animal metabolism studies are designed to obtain data on the absorption, distribution, metabolism, and excretion (ADME) of the test compound. A traditional approach to these ADME studies is to administer a ^{14}C-labeled analog of the parent material to the test species, which aids in data collection for each phase of the study. The greatest analytical challenge associated with these studies has been obtaining definitive metabolite identification and distribution data in the various tissues and excreta samples. Although the ^{14}C-labeled analog assists in locating many of the metabolites of interest, these samples generally possess very low absolute ^{14}C levels. This makes detecting the ^{14}C-labeled degradates by HPLC radiochemical analysis difficult.

One solution to this challenge is to search through all the accumulated MS data for compounds containing the correct isotopic cluster pattern, effectively using the MS as an isotope detector. However, this approach often yields limited data because the trace level unknowns present in the samples rarely provide sufficient signal for reliable isotopic cluster based filtering. One alternative approach is to utilize either an ion trap or triple quadrupole MS instrument operated in data-dependant MS/MS modes to provide structure data on these metabolites. This approach can often generate fragment ions or neutral losses which are characteristic of either the parent compound or predicted metabolites (i.e., conjugates). Unfortunately, neither of these instruments has the ability to provide the accurate mass data often critical for structure assignment. The accurate MS and MS/MS data generated using modern LC-TOFMS and LC-QqTOFMS instruments can greatly enhance the researcher's ability to propose a definitive structure for these metabolites, as discussed below.

The use of TOF and QqTOF mass spectrometers in toxicology studies has increased dramatically in recent years. This is due to several factors including improvements in mass resolution for the small molecules (mw < 800) typically encountered in toxicology studies, as well as advancements in instrument control and data-handling software. More user-friendly acquisition and data analysis software is now available, which allows additional users, who may have limited experience and expertise, to acquire and interpret experimental data from toxicology studies. Finally, the use of ADC-based signal processing has significantly enhanced the dynamic range of TOFMS detection, as discussed above. This ADC technology now provides detection linearity over 3–5 orders of magnitude, enabling the analyst to acquire useful (appropriate mass assignment and isotope ratios) full scan MS data for both small and large chromatographic peaks, often in a single chromatographic run. An additional benefit of this expanded dynamic range is realized when performing quantitative analysis. Although TDC-based TOFMS instruments previously had limited quantitative utility due to dynamic range limitations, the improved dynamic range of ADC-based TOFMS instruments has now expanded their use into quantitative applications, as discussed in the example below.

One recent application of LC/TOFMS was the analysis of hydroxyethyl valine (HEVal, Figure 10.7A) in F344 rat globin. A published quantitative GC/MS/MS method [22] was available for this analyte. However, modifying this method to LC/MS/MS promised to greatly simplify the sample preparation and analysis. Initial attempts were made to develop an LC/MS/MS method employing tandem MS with multiple-reaction monitoring (MRM). However, for this compound the MRM method showed an interference which negatively impacted the method detection limit (MDL). The method was then transferred to an LC-TOFMS instrument which employed ADC signal processing. Analysis was performed using full scan TOFMS acquisition. Using this approach, and applying narrow extracted ion windows (0.03 Da) for compound detection, the desired sensitivity was achieved without matrix interferences. Figure 10.7B (left panel) shows the detection of HEVal using this narrow extracted ion window approach on the Agilent LC-TOFMS. In addition,

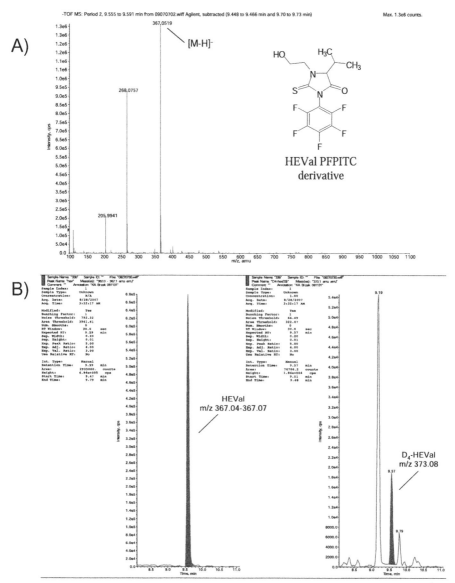

Figure 10.7.   A) Example of negative ion time-of-flight high resolution mass spectrum of hydroxyethyl valine (HEVal) derivatized with pentafluorophenyl isothiocyanate (PFPITC), B) Representative negative ion ESI-LC/TOFMS selected ion chromatograms of a derivatized sample extract sample #33, animal 5583, dose group 4, HEVal concentration 1.32 µg/mL (7.69 µg/g globin); D4-HEVal concentration 40 µg/mL. (Figure provided courtesy of Mrs. Kathy Brzak.)

this method required the use of stable isotope internal standard calibration. While the use of labeled internal standards can cause detector saturation in TDC-based TOFMS instruments, the increased dynamic range of the ADC-based instrument allowed the use of these standards (Figure 10.7B right panel), and provided quantitative detection of HEVal from 2 to 200 ng/mL with an $R^2 = 0.9963$.

10.5 BIOTECHNOLOGY APPLICATIONS OF LC-TOFMS

The combination of reversed-phase HPLC separations with electrospray ionization and mass spectral detection has revolutionized the biotechnology industry. Continual improvements in LC/MS interface designs, combined with powerful features for qualitative and quantitative mass spectral detection, have resulted in a widened scope of biotech applications for mass spectrometry. These improvements have coincided with breakthroughs in biochemistry, combinatorial chemistry, and molecular biology, dramatically accelerating industrial biotechnology discovery and product development. While LC/MS has influenced many areas of biotechnology, its most profound impact has been within protein analysis. Proteins and peptides are increasingly being developed as drugs for curing various diseases and in transgenic crops for controlling insects and protecting plants from herbicides. These proteins are often analyzed for their amino acid sequence, post-translational modifications, as well as their secondary, tertiary, and quaternary structures. Such information provides a scientifically sound basis for understanding the relationship of protein structure and function, for protein production process development and validation, and is essential for meeting regulatory requirements for registration of recombinantly expressed proteins.

Detailed characterization of proteins generally involves the identification of modified or altered amino acids in a single protein of interest. There have been many attempts to completely characterize a protein using a combination of proteolytic digestion and LC/MS, however the percent sequence coverage for many of these experiments has typically only been in the 25–50% range. Recently, technical advancements in chromatographic separations, including ultra performance liquid chromatography (UPLC) and comprehensive multidimensional liquid chromatography systems, coupled with tandem mass spectrometry, are significantly expanding the characterization and sequence coverage (75–99%) of post-translationally modified proteins [23–25]. These advancements are allowing the meaningful qualitative evaluation of product quality to a degree not possible with previous analytical approaches.

Although UPLC systems can achieve improved chromatographic resolution at flow rates of approximately 0.3 mL/min for small molecule analyses, one recent study [26] has shown that, for the separation of peptides, the optimal flow rate for these systems is only approx. 0.1 mL/min. When using a 2.1 mm internal diameter column, the van Deemter plot for a peptide of mass 1500 Da shows that the optimal linear velocity providing the best possible resolution is 0.33 mm/sec. In addition, higher column temperatures have been shown to improve chromatographic resolution of peaks during protein analysis [27, 28]. In order to demonstrate the increased

A)

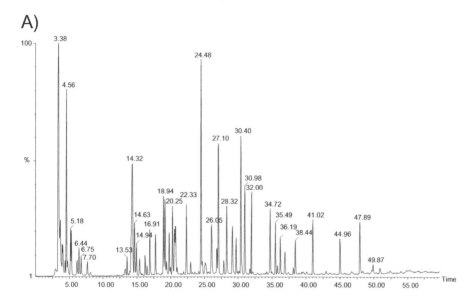

B)

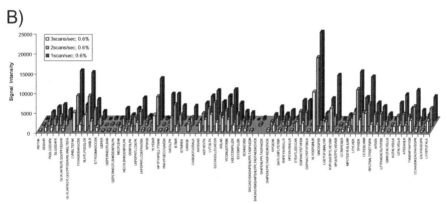

Figure 10.8.   (A) Total ion chromatogram of a 1 μg BSA tryptic digest on a 2.1 × 150 mm BEH C18 column. Integration was carried out using the Mass Lynx version 4.0 software, with automatic noise measurement, no smoothing, and peak detection by Apex Track Peak Integration with no threshold limits. (B) Ion intensity of the most intense charge state of a peptide is plotted against the peptide sequence for scan rates of 1, 2, and 3 scans/sec. (Figure provided courtesy of Dr. Krishna Kuppannan.)

performance of the UPLC/MS approach, bovine serum albumin (BSA) was studied using a Waters UPLC system coupled to a Q-TOF Micro instrument. BSA has been analyzed extensively; it is comprised of 607 amino acids with a theoretical molecular weight of 69,294 Da. Assuming no missed cleavages, a tryptic digest of BSA can produce a total of 82 peptides. For these studies, typical injection volumes were approximately 5 μL and the typical concentration used was 1 μg/5 μL. Figure 10.8A represents an optimized one-dimensional separation of BSA digest over a 60-minute analysis. Under these conditions, 54 BSA peptide peaks were detected, representing approximately 66% coverage, and the high resolution obtained using TOFMS provided sensitive detection combined with unambiguous assignment of peptide charge state.

In addition to developments in chromatographic separations, several mass spectrometer instrument parameters have been shown to influence the signal intensity and the quality of the resulting mass spectral data for proteins and peptides. One factor that influences peptide signal intensity is the rate of MS acquisition. In general, MS acquisition at higher scan rates produces lowered signal intensities. Conversely, slow scan rates may not provide sufficient data points to adequately characterize a chromatographic peak. In one example, scan rates of 1, 2 and 3 scans/sec were evaluated on a UPLC-QqTOFMS (Figure 10.8B). A comparison of these conditions indicates that at slower scan speeds, the total ion current increases, with acquisition at 1 scan/sec rate producing a signal intensity approximately four times that observed at 3 scans/sec. Further, there was no difference in overall peptide sequence coverage between the three MS acquisition rates, implying that on QqTOFMS systems, lower scan rates may prove better for peptide analysis, at least over the range investigated (approximately 500–3000 Da).

To further expand the ability to qualitatively and quantitatively characterize proteins, comprehensive two-dimensional liquid-phase (2DLC) separation systems have been coupled on-line to QqTOFMS detection. These 2DLC-MS systems can provide improved resolution of structural alterations to recombinant proteins. The improved resolution offered by the 2DLC separation of peptides was demonstrated by evaluating oxidized and unoxidized BSA. Figure 10.9 represents contour plots of the 2DLC-MS analysis for both BSA and oxidized BSA, where the "x" axis represents the different salt concentrations and the "y" axis represents the different retention times by RP-HPLC for the individual resolved peptides. By combining the results from all of the different salt/RP fractions, approximately 71 peaks were observed for both BSA and oxidized BSA, representing 87% sequence coverage. Figure 10.9C-D shows the elution of the oxidized and unoxidized fragment T75, TVMENFVAFVDK, in the 15 mM ammonium formate salt step. In Figure 10.9C, the unoxidized TVMENFVAFVDK was observed at a retention time of 137.95 min, which was subsequently confirmed by MS analysis (see inset). A new peak was

---

Figure 10.9. Two-dimensional contour plots comparing tryptic digest of (A) BSA and (B) oxidized BSA by 2DLC-MS at different ammonium formate concentrations. (C) Representative chromatogram of the elution of the unoxidized TVMENFVAFVDK (Fragment T75) in the 15 mM ammonium formate salt step. Inset shows the isotopic doubly charged mass ions of unoxidized TVMENFVAFVDK identified in the unoxidized BSA sample. (D) Representative chromatogram of the elution of the oxidized TVMENFVAFVDK L (Fragment T75) in the 10 mM formate salt step. Inset shows the isotopic doubly charged mass ions of oxidized TVMENFVAFVDK (Fragment T75, Met547) identified in the oxidized BSA sample.

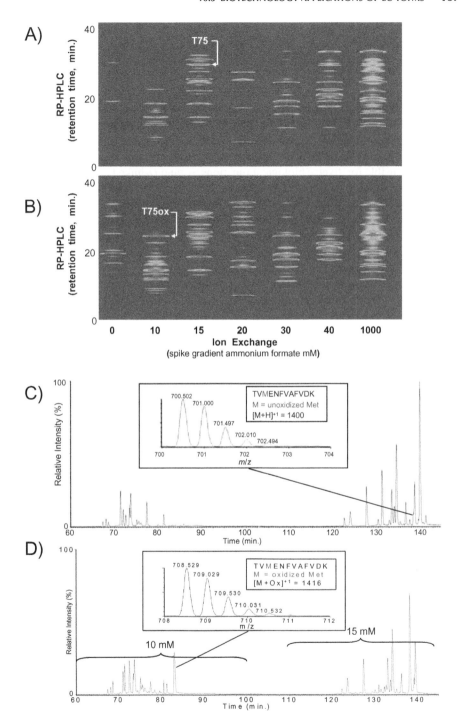

observed at 83.15 min in the oxidized BAS sample Figure 10.9D, which corresponds to the T75 oxidized fragment. The T75 fragment peak was identified as being oxidized (+16 u in mass), thus indicating that the peak is the oxidized Met547 form of BSA.

Another example where LC/TOFMS has had significant impact involves the analysis of post-translational modifications. Proteins can undergo a variety of post-translational modifications, including phosphorylation, oxidation, deamidation, proteolysis, and/or disulfide scrambling. In addition, glycans are often attached to proteins, particularly the *N*-linked oligosaccharide at highly conserved sites, which may be responsible for diverse biological functions. *N*- and *O*-linked oligosaccharide structures are attached to different sites on a glycoprotein (termed heterogeneity), and this heterogeneity often exists not only among these different sites, but also within the group of structures that occurs at each single site of glycosylation (termed microheterogeneity). This heterogeneity complicates the detailed characterization of the glycans present in glycoproteins, because it is very difficult to separate the oligosaccharide components for individual analysis [29].

High performance chromatography combined with electrospray ionization mass spectrometry has made glycoprotein analysis both feasible and practical. The combination of these technologies allows for the ability to ionize, selectively detect, and characterize covalently modified peptides among all the other peptides present in the proteolytic digest of a protein. For example, Figure 10.10A shows the ESI/MS total ion current chromatogram for the detection of *N*-linked oligosaccharides released from RNase B. The eluting peaks were assigned to individual RNase B oligosaccharides based on the corresponding mass spectra (Figure 10.10B). The known structures of the RNase B *N*-glycans are shown in Table 10.1.

Collectively, the key factor that facilitates detection, identification, and characterization of these complex glycans attached to glycoproteins is the use of HPLC separations combined with ESI and TOFMS. In addition, we have developed both relative and absolute quantitative methods to determine amounts of non-glycosylated, de-glycosylated, and glycosylated protein fragments, utilizing UPLC/UV/MS [25]. Other approaches for accurate quantitation of glycoproteins using HPLC/ESI/MS based methodology include isotopically substituted analogues of peptides of interest, which can be used as internal standards [30]. This is a promising avenue of development for quantitation of the glycosylation microheterogeneiety of a variety of glycoproteins of industrial interest.

10.6 CONCLUSIONS

In this chapter we have described several diverse industrial applications for time-of-flight based LC/MS. These can range from the identification of ultra-trace level metabolites generated when an agrochemical is exposed to an environmental system, to the identification of unknown biologically active compounds from natural sources. LC/TOFMS can be used to qualitatively identify metabolites generated in mammalian systems of toxicological interest, as well as to quantify their levels. LC/TOFMS also plays a critical role in both the identification and detailed

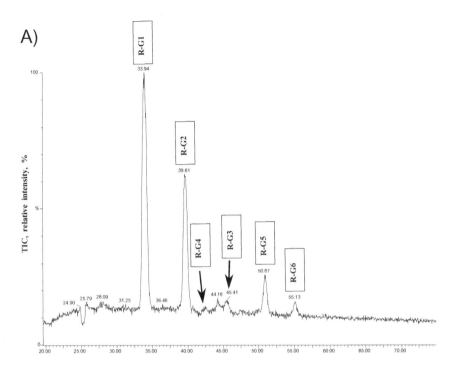

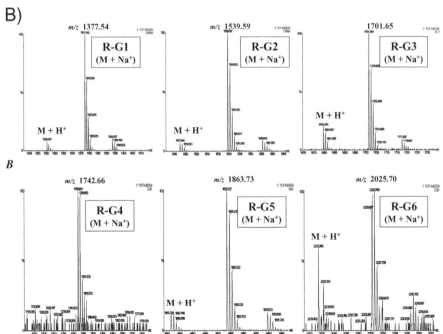

Figure 10.10.   (A) HPLC (GlycoSep N column) with ESI-MS detection of *N*-linked oligosaccharides released from RNase B. (B) Mass spectra corresponding to the LC peaks that confirm the identity of the oligosaccharides. Refer to Table 10.1 for oligosaccharide structures. (Figure provided courtesy of Dr. Anton Karnoup.)

TABLE 10.1. Structures of Oligosaccharides (Glycans) Released by RNase B Standard Glycoprotein

Glycan ID	Structure of Glycan[a]	Theor. m/z[b]	Observed m/z[c]
R-G1	(Man)$_5$(GlcNAc)$_2$	1377.501	1377.542
R-G2	(Man)$_6$(GlcNAc)$_2$	1539.562	1539.592
R-G3	(Man)$_7$(GlcNAc)$_2$	1701.615	1701.651
R-G4	(Gal)(Man)$_5$(GlcNAc)$_3$	1742.641	1742.661
R-G5	(Man)$_8$(GlcNAc)$_2$	1863.667	1863.727
R-G6	(Man)$_9$(GlcNAc)$_2$	2025.720	2025.697

characterization of the proteins generated in industrial biotechnology. Through the combination of developments in high performance separations, coupled with the high speed and high resolution detection provided by LC/TOFMS systems, detailed information on the primary structure, post-translational modifications, and even secondary, tertiary, and quaternary structure of proteins can be obtained. The exceptional combination of acquisition speed, full scan sensitivity, high resolution, and high mass accuracy provided by LC/TOFMS systems has made them an indispensable tool for a wide variety of industrial applications. Recent advances in TOFMS detection, including ADC-based detectors, have also opened up new applications for LC/TOFMS in quantitative analyses. This unique combination of features promises to ensure a rich future of growth for LC/TOFMS in industrial applications.

REFERENCES

1. Balogh, M.P., The commercialization of LC-MS during 1987–1997: A review of ten successful years, *LC-GC*, **1998**, 16(2), 135–136, 138, 140, 142, 144.

2. Niessen, W.M.A., Advances in instrumentation in liquid chromatography-mass spectrometry and related liquid-introduction techniques, *J Chromatogr A*, **1998**, 794, 407–435.

3. Morris, H.R.; Paxton, T.; Dell, A.; Langhorne, J.; Berg, M.; Bordoli, R.S.; Hoyes, J.; Bateman, R. H., High sensitivity collisionally-activated decomposition tandem mass spectrometry on a novel quadrupole/orthogonal-acceleration time-of-flight mass spectrometer, *Rapid Commun Mass Spectrom*, **1996**, 10, 889–896.

4. Shevchenko, A.; Chernushevich, I.; Ens, W.; Standing, K.G.; Thomson, B.; Wilm, M.; Mann, M., Rapid "de novo" peptide sequencing by a combination of nanoelectrospray, isotopic labeling and a quadrupole/time-of-flight mass spectrometer, *Rapid Commun Mass Spectrom*, **1997**, 11, 1015–1024.

5. Eckers, C.; Haskins, N.; Langridge, J., The use of liquid chromatography combined with a quadrupole time-of-flight analyzer for the identification of trace impurities in drug substance, *Rapid Commun Mass Spectrom*, **1997**, 11, 1916–1922.

6. Eckers, C.; Wolff, J.-C.; Haskins, N.J.; Sage, A.B.; Giles, K.; Bateman, R., Accurate mass liquid chromatography/mass spectrometry on orthogonal acceleration time-of-flight mass analyzers using switching between separate sample and reference sprays. 1. Proof of concept, *Anal. Chem.*, **2000**, 72, 3683–3688.

7. Wolff, J.-C.; Eckers, C.; Sage, A.B.; Giles, K.; Bateman, R., Accurate mass liquid chromatography/ mass spectrometry on quadrupole orthogonal acceleration time-of-flight mass analyzers using switching between separate sample and reference sprays. 2. Applications using the dual-electrospray ion source, *Anal. Chem.*, **2001**, 73, 2605–2612.

8. Fjeldsted, J., *Time-of-Flight Mass Spectrometry*, Agilent Technologies Technical Overview #59890-0373EN.

9. McIntyre, D., *Advantages of Wide Dynamic Range on an Orthogonal Acceleration Time-of-Flight Mass Spectrometer*, Agilent Technologies Technical Overview #5989-1728EN.

10. Ferrer, I.; Thurman, E.M.; Garcia-Reyes, J.F.; Fernandez-Alba, A., Identification and quantitation of pesticides in vegetables by liquid chromatography time-of-flight mass spectrometry, *Trends Anal. Chem.*, **2005**, 24, 671–682.

11. Thurman, E.M.; Ferrer, I.; Zweigenbaum, J.A., High-resolution and accurate mass analysis of xenobiotics in food. *Anal. Chem.*, **2006**, 78, 6702–6708.

12. Constant, H.L.; Beecher, C.W., A method for the dereplication of natural product extracts using electrospray HPLC/MS, *Nat Prod Lett*, **1995**, 6, 193–196.

13. Kebarle, P.; Tang, L., From ions in solution to ions in the gas phase—the mechanism of electrospray mass spectrometry, *Anal Chem*, **1993**, 65, 972A–986A.

14. Gilbert, J.R.; Lewer, P.; Carr, A.W.; Snipes, C.E.; Balcer, J.L., Gerwick, W., Natural product dereplication and structural elucidation using LC/MSn combined with accurate mass LC/MS and LC/MS/MS, Proceedings of the 47th ASMS Conference on Mass Spectrometry and Allied Topics, Dallas, Texas, June 13–17, **1999**.

15. Potterat, O.; Wagner, K.; Haag, H., Liquid chromatography-electrospray time-of-flight mass spectrometry for on-line accurate mass determination and identification of cyclodepsipeptides in a crude extract of the fungus metarrhizium anisopliae, *J Chromatogr A*, **2000**, 872, 85–90.

16. Dictionary of Natural Products on DVD; **2007**; Chapman and Hall/CRC.

17. Gilbert, J.R.; Lewer, P.; Duebelbeis, D.O.; Carr, A.W.; Snipes, C.E.; Williamson, R.T., Identification of biologically-active compounds from nature using LC-MS in Mass Spectrometry, LC/MS/MS and TOF/MS: Analysis of Emerging Contaminants, ACS Symposium vol. 80, Imma Ferrer and E.M. Thurman, Eds., Oxford University Press and the American Chemical Society (2003).

18. Marfey, P., Determination of D-amino acids. II. Use of a bifunctional reagent, 1,5-difluoro-2,4-dinitrobenzene, *Carlsberg Res. Commun.*, **1984**, 49, 591–596.

19. Nielsen, K.F.; Smedsgaard, J., Fungal metabolite screening: database of 474 mycotoxins and fungal metabolites for dereplication by standardized liquid chromatography-UV-mass spectrometry methodology, *J Chromatogr A*, **2003**, 1002, 111–136.

20. Balcer, J.L; Gilbert, J.R.; Linder, S.J.; Magnussen, J.D.; Johnson, P.J.; Krieger, M., Identification of novel plant metabolites using accurate mass, MS/MS data, nanospray technology, and unique isotope pattern recognition, Proceedings of the 55th ASMS Conference on Mass Spectrometry and Allied Topics, Indianapolis, IN, June 3–7, **2006**.

21. Hager, J. W., A new linear ion trap mass spectrometer, *Rapid Commun. Mass Spectrom.*, **2002**, 16, 512–526.

22. Tornqvist, M.; Mowrer, J.; Jensen, S.; Ehrenberg, L., Monitoring of environmental cancer initiators through hemoglobin adducts by a modified Edman degradation method., *Anal. Biochem*, **1986**, 154, 255–266.

23. Gilar, M.; Olivova, P.; Daly, E.; Gebler, J.C., Orthogonality of Separation in Two-Dimensional Liquid Chromatography, *Anal. Chem.*, **2005**, 77, 6426–6434.

24. Kajdan, T.; Cortes, H.; Kuppannan, K.; Young, S.A., Development of a comprehensive multidimensional liquid chromatography system with tandem mass spectrometry detection for detailed characterization of recombinant proteins, *J Chromatogr A*, **2008**, 1189, 183–195.

25. Karnoup, A.S.; Kuppannan, K.; Young, S.A., A novel HPLC-UV-MS method for quantitative analysis of protein glycosylation, *J Chromatogr B*, **2007**, 859, 178–191.

26. Mazzeo, J.; Wheat, T.; Gllece-Castro, B.; Lu, Z., Next generation peptide mapping with ultra performance liquid chromatography, *BioPharm. International*, **2006**, 19, 22–23.

27. Yang, X.; Ma, L.; Carr, P.W., High temperature fast chromatography of proteins using a silica-based stationary phase with greatly enhanced low pH stability, *J. Chromatogr A*, **2005**, 1079, 213–220.

28. Dillon, T.M.; Bondarenko, P.V.; Rehder, D.S.; Pipes, G.D.; Kleemann, G.R.; Ricci, M.S., Optimization of a reversed-phase high-performance liquid chromatography/mass spectrometry method for characterizing recombinant antibody heterogeneity and stability, *J. Chromatogr A*, **2006**, 1120, 112–120.

29. Selttineri, C.A.; Burllngame, A.L. In Z. El Rassi (Editor), Carbohydrate Analysis: High Performance Liquid Chromatography and Capillary Electrophoresis. Elsevier, Amsterdam, **1995**, p. 447.

30. Kirkpatrick, D.S.; Gerber, S.A.; Gygi, S.P., The absolute quantification strategy: a general procedure for the quantification of proteins and post-translational modifications, *Methods*, **2005**, 35, 265–273.

CHAPTER 11

COMPREHENSIVE TOXICOLOGICAL AND FORENSIC DRUG SCREENING BY LC/TOF-MS

Ilkka Ojanperä, Anna Pelander, and Suvi Ojanperä

Department of Forensic Medicine, University of Helsinki, Helsinki, Finland

MONITORING SMALL molecular weight organic compounds is important in pharmaceutical research, environmental hygiene, food safety, doping control, and forensic science and toxicology. Among the several mass spectrometric techniques available, new-generation LC/TOF-MS is gaining increasing attention. The present chapter describes the performance of the authors' LC/TOF-MS-based screening approach in forensic applications. The method uses the principle of accurate mass measurement combined with a reverse search based on a large target database of exact monoisotopic masses. Dedicated software is utilized to perform mass scale calibration of the data, create extracted ion chromatograms (EIC) in a 0.002 *m/z* mass window for the protonated molecule of each molecular formula included in the list, apply peak detection and identification criteria according to mass accuracy, isotopic pattern, area and retention time, if available, and finally create an MS Excel-based result report. The LC/TOF-MS method is applied to the comprehensive screening analysis of drugs and pharmaceuticals in urine, vitreous humor, and hair samples in a forensic toxicology context. Moreover, LC/TOF-MS has been shown to be a viable technique for the tentative identification of designer drugs in seized samples without using reference standards.

11.1 INTRODUCTION

Many diverse areas of society benefit from screening analysis of small molecular weight organic compounds. These areas include research, monitoring, and enforcement aspects and such analytical tasks as screening for combinatorial libraries at the

Liquid Chromatography Time-of-Flight Mass Spectrometry: Principles, Tools, and Applications for Accurate Mass Analysis, Edited by Imma Ferrer and E. Michael Thurman

early stages of drug discovery [1], detection of pollutants in environmental hygiene [2], monitoring of pesticides to improve food safety [3], analysis of drugs and medicines in the context of doping control [4], and screening for medicinal substances, drugs and poisons within forensic and clinical toxicology [5]. Broad-spectrum screening procedures, comprising hundreds of analytes of known or partly known structure, have conventionally relied on the combined use of a multitude of analytical techniques. This setting is changing, however, due to the introduction of powerful liquid chromatography-mass spectrometry (LC/MS) based technologies. While quadrupole LC/MS is still predominant, techniques for accurate mass measurement, especially orthogonal acceleration time-of-flight mass spectrometry (TOF-MS), have gained wide acceptance in small molecule analysis over the last decade [6]. The benefits of modern TOF-MS analyzers include high speed, good sensitivity, sufficient resolution, and in particular, a high mass accuracy comparable to that of much more expensive accurate mass instruments [7]. These features, together with the method's robustness and ease of use, make LC/TOF-MS a particularly competitive alternative for use as the main screening instrument of toxicology and forensic science laboratories.

11.2   COMPREHENSIVE TOXICOLOGICAL AND FORENSIC DRUG SCREENING

Broad-spectrum screening analysis plays a prominent role in various fields of toxicology and forensic science [8]. Optimal performance is required in post-mortem toxicology studies related to cause-of-death investigations, and laboratories should therefore be able to include 500–1000 medicinal substances, illicit drugs, and poisons in the routine qualitative analyses of body fluids. Similar demands apply to clinical forensic toxicology, which comprises the investigation of the toxicological aspects of violent crime, chemical exposure, child welfare, and drug use and trafficking. In clinical toxicology at the emergency department, a comprehensive drug screening service would facilitate medical treatment, but a limited analysis is usually necessary due to time restrictions. Controlling driving under the influence of drugs is limited to substances that impair driving performance, while doping control in sports concerns substances on the prohibited list of the World Anti-Doping Agency. Drug testing at the workplace, in prisons, and at schools is usually restricted to illicit drugs.

The standard handbook on analytical toxicology by Baselt [9] covers nearly 600 relevant organic toxicants, of which about 80% contain nitrogen. Major drugs of abuse that do not contain nitrogen are few; they include, e.g., gamma-hydroxybutyrate and cannabinoids. Sample preparation for instrumental analysis usually involves division of the analytes into basic and acidic fractions. The basic fraction contains the most important toxicologically relevant classes of drugs and pharmaceuticals, especially substances that act on the central nervous system (CNS), such as amphetamines, antidepressants, antipsychotics, benzodiazepines, and opioids. Features of a successful CNS-active agent include log P < 5 (mean 2.5), molecular weight <450 and neutral or basic character with a pK_a of 7.5–10.5 [10].

Previously, simultaneous screening and quantification has been feasible only if quantitative calibration of the method is required on a weekly or monthly basis, such as when using gas chromatography (GC) with nitrogen-selective detection [11] or when using LC with diode-array UV detection [12]. Analysis techniques that necessitate quantitative calibration in each sequence of runs, such as LC/MS, have typically been employed for only 10–20 compounds. Recently, a quantitative method was published for up to 100 pesticides in food using a triple quadrupole technique (LC/MS/MS) [13], and this approach will obviously have implications for drug analysis, too.

In qualitative drug screening, GC/MS-based techniques have traditionally provided an unsurpassed performance. Several extensive electron ionization MS libraries are commercially available to facilitate broad-scale screening for organic small molecular weight compounds, including drugs, medicines, poisons, and pesticides. Recently, LC/MS/MS techniques, applying triple quadrupole or hybrid triple quadrupole linear ion trap technology, have also proven successful in comprehensive drug screening [14]. These procedures use multiple reaction monitoring mode (MRM), (enhanced) product ion scan, or a combination of the two. The reproducibility of LC/MS/MS spectra has often been questioned. However, methods to standardize libraries of MS/MS spectra from triple quadrupole [15] and hybrid instruments [16] and from several types of LC/MS/MS instruments [17] have recently been published.

A typical LC-MS drug screening procedure is based on chromatographic separation on a C-18 reversed phase column using a gradient with acetonitrile and aqueous ammonium formate as the mobile phase. Instrumentation involves triple quadrupole LC/MS/MS applying electrospray ionization in the positive mode. The mass analyzer is used in the MRM or product ion scan mode for quantification or qualitative screening, respectively [18]. When compared to LC/TOF-MS techniques, the disadvantages of the above-mentioned quadrupole LC-MS techniques include their limited sensitivity in the full scan mode and the need for a reference substance for identification.

11.3 THE LEVEL OF LC/TOF-MS PERFORMANCE REQUIRED FOR SCREENING

New generation TOF-MS instruments allow measurements with a mass accuracy better than 5 ppm, which has been found to be sufficient for drug screening [19], and for mass resolution exceeding 10,000 FWHM. An equally important property of TOF-MS is the improved ion-abundance dynamic range for accurate mass measurement achieved by using technology based on an analogue-to-digital converter (ADC). A wide dynamic range is a prerequisite for an automated target screening procedure as it allows accurate mass measurements on a routine basis without tedious optimization procedures. Instruments equipped with ADC have been successfully used in recent screening applications for medicinal substances [20] and pesticides [21]. A unique feature for generating molecular formulae is the SigmaFit™ algorithm (Bruker Daltonics, Bremen, Germany), which provides an exact numerical comparison between theoretical and measured isotopic patterns as an additional identification tool for accurate mass determinations. It is now possible to measure isotopic ratios with a 2% intensity precision and to use this feature with the application of the SigmaFit™ calculation in the software. In comprehensive screening, the size and scope of the target database should be carefully designed for the particular application in question—such as for drugs, pesticides or steroids—in order to avoid misleading false positive findings. In all types of screening, dedicated software is necessary to condense the vast amount of data produced and report the analytical results in a reliable and straightforward manner.

11.4   URINE DRUG SCREENING BY LC/TOF-MS

Urine is the most important material for qualitative toxicological drug screening since the substances and/or their metabolites are present there in higher concentrations than in the blood. The time-window for detecting illicit drug use is also larger than with blood, being often several days. In 2001, our research group introduced a novel target screening concept for drugs and medicinal substances in urine, using formula-based identification by LC/TOF-MS [22], and the method has since been improved and validated [23]. In addition to forensic and clinical toxicology, the method has been found useful for doping control [24]. The approach is original in its use of the principle of accurate mass measurement for high-throughput toxicological drug screening in a biological matrix combined with a reverse search based on a target database of exact monoisotopic masses. Entries in the large target database, representing the elemental formulae of reference substances and their metabolites, are compared with the measured masses for protonated molecules MH$^+$.

The method involves sample preparation based on cation exchange/C-4 mixed-mode solid-phase extraction of urine samples hydrolyzed with a glucuronidase enzyme [20]. The procedure allows extraction of basic and amphoteric compounds of a wide polarity range, yet provides a reasonably clean background. Chromatographic separation is performed on a Luna C-18(2) $100 \times 2 \, mm$ ($3 \, \mu m$) (Phenomenex, Torrance, CA) column using a gradient with $5 \, mM$ ammonium acetate in 0.1% formic acid and acetonitrile as the mobile phase. Electrospray ionization TOF-MS is operated in the positive mode. Currently, the mass analyzer in our use is a Bruker micrOTOF, which represents high performance in terms of mass accuracy and the dynamic range available for accurate mass measurement. In addition to accurate mass measurement, identification relies on isotopic pattern fit, retention time (if a reference substance is available), and known metabolite patterns in urine. In a study involving autopsy urine samples, the mean and median of mass error absolute values for true-positive findings were $2.5 \, ppm$ and $2.2 \, ppm$, respectively, corresponding to $0.65 \, mDa$ and $0.60 \, mDa$ [20]. SigmaFit™ was found to reveal true-positive findings and produce on average 12% less false-positive entries than using accurate mass only in target analysis [20].

Calibration of the mass analyzer is essential in order to maintain a high level of mass accuracy. In the present procedure, instrument calibration is performed externally prior to each sequence with sodium formate solution, consisting of $10 \, mM$

sodium hydroxide in isopropanol-0.2% formic acid. Automated internal mass scale calibration of individual samples is performed by injecting the calibrant at the beginning and at the end of each run via a 6-port divert valve. Calibration is performed post-run based on the first calibrant injection, while the same calibrant at the end of run is for manual verification of calibration stability. The calibration ions in the six-point post-run internal mass scale calibration comprise the following sodium formate/acetate clusters: $C_2H_2Na_3O_4$ m/z 158.96407, $C_4H_5Na_4O_6$ m/z 240.96714, $C_5H_5Na_6O_{10}$ m/z 362.92634, $C_6H_7Na_6O_{10}$ m/z 376.94200, $C_6H_6Na_7O_{12}$ m/z 430.91377, and $C_7H_8Na_7O_{12}$ m/z 444.92942. Room temperature changes of no more than $\pm 2\,°C$ are a prerequisite for optimal performance. The retention times require recalibration every six months.

Advanced software features are indispensable for high-throughput analysis. TargetAnalysis 1.1 (Bruker Daltonics) software has been developed based on many years of experience in our laboratory in conducting toxicological drug screening by LC/TOF-MS. TargetAnalysis is a new software that includes tools for various screening and target analysis applications. Our current in-house database for toxicological urine screening includes 815 masses comprising a wide variety of medicinal substances, drugs of abuse, metabolites, and designer drugs. Retention time information is available for approximately 50% of these compounds. TargetAnalysis software performs the mass scale calibration of the data; creates extracted ion chromatograms (EIC) in a $0.002\,m/z$ mass window for the MH^+ of each molecular formula included in the list; applies peak detection and identification criteria according to mass accuracy, isotopic pattern (SigmaFit™), area, and retention time, if available; and finally creates an MS Excel-based result report. The two-level criteria for mass accuracy, SigmaFit™ and retention time error, are 5 and 8 ppm, 0.03 and 0.05, and 0.1 and 0.3 min, respectively. Score 3 (positively identified, highlighted in dark gray) is given for an entry with better identification parameters than 5 ppm mass accuracy, 0.03 SigmaFit™, and 0.1 min retention time. Score 2 (probably identified, highlighted in light gray) is given for an entry with any of the three parameters between the two levels. Score 1 (tentatively identified, not highlighted) is given for compounds without a reference retention time. An area cut-off of 50,000 is applied to all identification levels.

Figure 11.1 shows a total ion chromatogram (TIC) for a post-mortem urine sample, a set of EICs, and a report created by the TargetAnalysis software. The report identifies 13 parent compounds and the internal standard dibenzepin, together with tentative metabolite identification for three carbamazepine metabolites, one cotinine metabolite, and one nitrazepam metabolite. Cotinine, 7-aminonitrazepam, and the tramadol metabolites are considered as the parent compounds since the reference standards were available for those. The substances identified represent a wide range of polarity and pK_a properties, despite the complexity of the urine matrix. This demonstrates well the advantages of this screening approach. The entries in the report for clobazam and norclobazam are false positive findings as the corresponding masses are assigned for temazepam and oxazepam, respectively. In addition, the mass assigned for nortramadol and O-desmethyltramadol is falsely reported as dinorvenlafaxine. These false positive findings demonstrate the fact that targets with identical molecular formulae can result in multiple identifications. Moreover, the

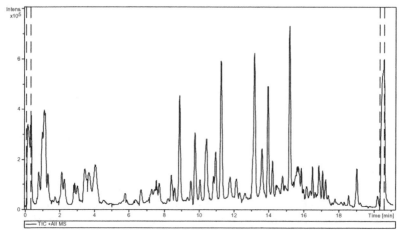

Case 944/07 urine TIC.

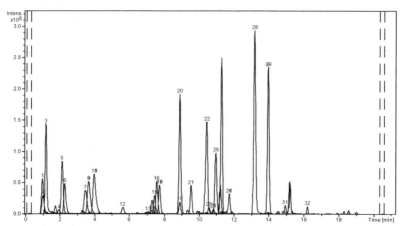

Case 944 /07 urine EICs.

Found	Compound Name	Reg.No.	Mol.Formula	PMI	d RT [min]	Err [ppm]	Err [mDa]	Sigma	Area	Intens.	RT exp.[min]	RT meas.[min]	m/z calc.	m/z meas.
3	OXAZEPAM	155	C 15 H 11 Cl 1 N 2 O 2	[M+H]+	-0.02	1.8	0.5	0.030	1661702	233850	13.92	13.94	287.0582	287.0577
3	TEMAZEPAM	1343	C16H13Cl1N2O2	[M+H]+	0.01	2.4	0.7	0.023	2399026	339372	15.17	15.16	301.0738	301.0731
3	CARBAMAZEPINE	1341	C 15 H 12 N 2 O 1	[M+H]+	-0.02	2.9	0.7	0.015	2407139	292564	13.15	13.17	237.1022	237.1015
1	CARBAMAZEPINE-10-11-EPOXIDE	1542	C 15 H 12 N 2 O 2	[M+H]+	-8.88	0.8	0.2	0.009	144740	19996	0.00	8.88	253.0972	253.09 70
1	CARBAMAZEPINE-10-11-EPOXIDE	1542	C 15 H 12 N 2 O 2	[M+H]+	-10.43	2.3	0.6	0.019	1364112	146712	0.00	10.43	253.0972	25 3.0966
1	CARBAMAZEPINE-10-11-EPOXIDE	1542	C 15 H 12 N 2 O 2	[M+H]+	-10.95	1.4	0.4	0.016	753746	97376	0.00	10.95	253.0972	253.0968
1	CARBAMAZEPINE-10-11-EPOXIDE	1542	C 15 H 12 N 2 O 2	[M+H]+	-11.72	0.7	-0.2	0.016	248835	32168	0.00	11.72	253.0972	253 .0973
1	10-11-DIHYDROXYCARBAMAZEPINE	1543	C 15 H 14 N 2 O 3	[M+H]+	-8.88	1.3	0.4	0.015	1418302	190904	0.00	8.88	271.1077	271.1074
1	IMINOSTILBENE	1544	C 14 H 11 N 1	[M+H]+	-7.30	1.1	0.2	0.011	52020	5423	0.00	7.30	194.0964	194.0962
2	CLOBAZAM	3831	C16H13Cl1N2O2	[M+H]+	0.23	2.4	0.7	0.023	2399026	339372	15.17	15.16	301.0738	301.0731
1	NORCLOBAZAM	3832	C 15 H 11 Cl 1 N 2 O 2	[M+H]+	-13.94	1.8	0.5	0.030	1661702	233850	0.00	13.94	287.0582	287.0577
2	COTININE	4932	C 10 H 12 N 2 O 1	[M+H]+	-0.04	0.3	0.1	0.013	1043060	143682	1.32	1.22	177.1022	177.1026
1	HYDROXYCOTININE	4933	C 10 H 12 N 2 O 2	[M+H]+	-1.02	2.1	-0.4	0.014	459026	55996	0.00	1.02	193.0972	193.0 976
3	PARACETAMOL	5022	C 8 H 9 N 1 O 2	[M+H]+	0.03	4.7	-0.7	0.008	733989	84406	2.19	2.16	152.0706	152.0713
1	PHENYLETHYLMALONAMIDE	5742	C 11 H 14 N 2 O 2	[M+H]+	-1.98	1.4	-0.3	0.006	60484	6677	0.00	1.98	207.1128	207.113 1
1	4-HYDROXYALPRENOLOL	7632	C 15 H 23 N 1 O 3	[M+H]+	-5.65	0.6	0.2	0.018	101185	10815	0.00	5.65	266.1751	266.1749
1	NORALPRENOLOL	7633	C 14 H 21 N 1 O 2	[M+H]+	-3.98	0.6	0.2	0.016	795556	63082	0.00	3.98	236.1645	236.1644
2	7-AMINONITRAZEPAM	7852	C 15 H 13 N 3 O 1	[M+H]+	0.00	0.3	-0.1	0.003	460024	37643	3.49	3.49	252.1131	252.1132
1	7-ACETAMIDONITRAZEPAM	7853	C 17 H 15 N 3 O 2	[M+H]+	-7.44	0.3	0.1	0.008	232158	29366	0.00	7.44	294.1237	294.1236
3	TRAMADOL	8861	C 16 H 25 N 1 O 2	[M+H]+	-0.01	0.5	0.1	0.014	429616	52308	7.56	7.57	264.1958	264.19 57
3	O-DESMETHYLTRAMADOL	8862	C 15 H 23 N 1 O 2	[M+H]+	-0.04	0.3	0.1	0.010	655433	51990	3.63	3.67	250.1802	250.1801
2	NORTRAMADOL	8863	C 15 H 23 N 1 O 2	[M+H]+	-0.04	0.2	0.1	0.001	369233	45969	7.69	7.73	250.1802	250.1801
3	O-DESMETHYLNORTRAMADOL	8864	C 14 H 21 N 1 O 2	[M+H]+	-0.02	0.6	0.2	0.016	795556	63082	3.95	3.98	236.1645	236.16 44
1	NORMEDAZEPAM	9466	C 15 H 13 Cl 1 N 2	[M+H]+	-10.56	0.6	0.1	0.003	90681	11034	0.00	10.56	257.0840	257.0839
1	DINORVENLAFAXINE	9844	C 15 H 23 N 1 O 2	[M+H]+	-3.67	0.3	0.1	0.010	655433	51990	0.00	3.67	250.1802	250.1801
1	DINORVENLAFAXINE	9844	C 15 H 23 N 1 O 2	[M+H]+	-7.73	0.2	0.1	0.001	369233	45969	0.00	7.73	250.1802	250.1801
1	WARFARIN	11121	C 19 H 16 O 4	[M+H]+	-0.03	0.9	0.3	0.002	71956	12156	16.14	16.17	309.1121	309 .1119
2	OXCARBAZEPINE	15321	C 15 H 12 N 2 O 2	[M+H]+	-0.17	0.7	-0.2	0.016	248835	32168	11.58	11.72	253.0972	253.0973
5	LEVETIRACETAM	16521	C 8 H 14 N 2 O 2	[M+H]+	0.02	2.5	-0.4	0.008	440504	48349	2.34	2.33	171.1128	171.1132
3	DIBENZEPIN	99909	C 18 H 21 N 3 O 1	[M+H]+	-0.02	0.9	0.3	0.012	329244	45718	9.50	9.52	296.1787	296.1735
3	BISOPROLOL	99989	C18H31N1O4	[M+H]+	-0.01	2.9	0.9	0.025	1269539	193729	9.75	9.76	326.2326	326.2317

Case 944/07 urine results report.

role of interpretation of the results with a possible request for confirmation analysis is emphasized.

11.5   DRUG SCREENING IN ALTERNATIVE MATRICES BY LC/TOF-MS

Although urine is an important sample material for toxicological drug screening, other materials, such as hair, saliva, and vitreous humor, are also commonly used [25]. In post-mortem toxicology, vitreous humor is a good alternative to urine, which is not always available. Saliva is a practical alternative, especially in the control of driving under the influence of drugs since the sampling procedure is noninvasive. Drugs incorporated in the hair can reveal even a single dose months afterward. To date, LC/TOF-MS has only occasionally been applied to alternative matrices, e.g., to detect cannabis in saliva [26] and drugs of abuse in hair [27], but no comprehensive screening methods have been published. The following examples show the viability of the present technique in these important areas.

Figure 11.2 shows a TIC, a set of EICs and a report created by the Target-Analysis software for a vitreous humor sample related to the same case as Figure 11.1. The sample preparation procedure for vitreous humor was identical to that used for urine [20], but excluded the enzyme hydrolysis step. The TIC shows a much cleaner background than found in the urine sample, and the results report is accordingly much more straightforward. This example demonstrates the applicability of vitreous humor as an alternative matrix to urine. As a viscous liquid, the material is easy to handle compared to fat-containing liver, which is a common material in post-mortem toxicology. Moreover, even relatively hydrophilic analytes can be readily detected in vitreous humor, which is not the case with liver.

Figure 11.3 shows a TIC for a hair sample, and a set of EICs and a report created by the TargetAnalysis software. The sample preparation procedure for hair included alkaline digestion in sodium hydroxide solution prior to the same solid-phase extraction method that was used for urine and vitreous humor analysis. In

---

Figure 11.1.   Total ion chromatogram (TIC), extracted ion chromatograms (EIC) and results report (TargetAnalysis 1.1) for an autopsy urine sample analyzed by the LC/TOF-MS method with an 815 compound database (see text). The two-level criteria for mass accuracy, SigmaFit™ and retention time error are 5 and 8 ppm, 0.03 and 0.05, and 0.1 and 0.3 min, respectively. Score 3 (positively identified, dark gray highlighted) is given for an entry with identification parameters better than 5 ppm mass accuracy, 0.03 SigmaFit™, and 0.1 min retention time. Score 2 (probably identified, light gray highlighted) is given for an entry with any of the three parameters between the two levels. Score 1 (tentatively identified, not highlighted) is given for compounds without a reference retention time. The five-digit register number refers to parent drugs and their metabolites as follows: the three first digits from the left refer to the drug group (1–999), the fourth digit refers to the number of compounds in the group (1–9), and the fifth digit assigns the compound in a group (1–9). Dibenzepin is an internal standard.

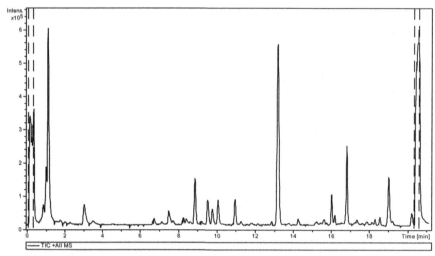

Case 944/07 vitreous humor TIC.

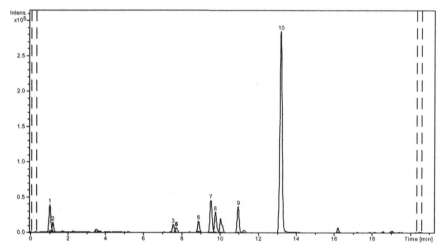

Case 944/07 EICs.

Found	Compound Name	Reg.No.	Mol.Formula	PMI	d RT [min]	Err [ppm]	Err [mDa]	Sigma	Area	Intens.	RT exp.[min]	RT meas.[min]	m/z calc.	m/z meas.
3	CARBAMAZEPINE	1541	C 15 H 12 N 2 O 1	[M+H]+	-0.01	2.5	0.6	0.023	2466422	284277	13.15	13.16	237.1022	237.1017
1	CARBAMAZEPINE-10-11-EPOXIDE	1542	C 15 H 12 N 2 O 2	[M+H]+	-10.95	2.8	0.7	0.012	254280	36904	0.00	10.95	253.0972	253.0964
1	10-11-DIHYDROXYCARBAMAZEPINE	1543	C 15 H 14 N 2 O 3	[M+H]+	-8.88	2.2	-0.6	0.012	113242	16100	0.00	8.88	271.1077	271.1083
3	COTININE	4932	C 10 H 12 N 2 O 1	[M+H]+	0.99	3.6	-0.6	0.004	92638	14274	1.32	1.23	177.1022	177.1029
3	TRAMADOL	8861	C 16 H 25 N 1 O 2	[M+H]+	-0.02	0.0	0.0	0.010	98961	11380	7.56	7.58	264.1958	264.1958
3	NORTRAMADOL	8863	C 15 H 23 N 1 O 2	[M+H]+	-0.04	0.3	-0.1	0.015	50085	6491	7.69	7.73	250.1802	250.1802
1	DINORVENLAFAXINE	9844	C 15 H 23 N 1 O 2	[M+H]+	-7.73	0.3	-0.1	0.015	50085	6491	0.00	7.73	250.1802	250.1802
3	WARFARIN	11121	C19H16O4	[M+H]+	0.02	2.2	0.7	0.012	74786	12185	16.16	16.14	309.1121	309.1114
3	DIBENZEPIN	99999	C 18 H 21 N 3 O 1	[M+H]+	-0.02	0.5	0.1	0.011	343657	45356	9.50	9.52	296.1757	296.1756
3	BISOPROLOL	99999	C 18 H 31 N 1 O 4	[M+H]+	-0.01	0.1	0.0	0.016	218549	28725	9.75	9.76	326.2326	326.2326

Case 944/07 vitreous humor results report.

Figure 11.2. TIC, EICs, and results report for a vitreous humor sample of the same case as in Figure 1 analyzed by the LC/MS-TOF method (see text).

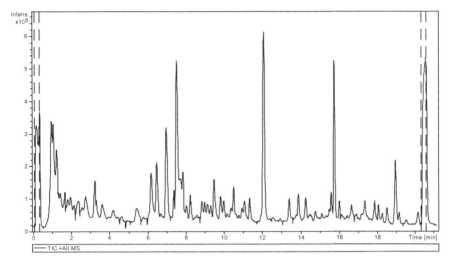

4359/07 hair TIC.

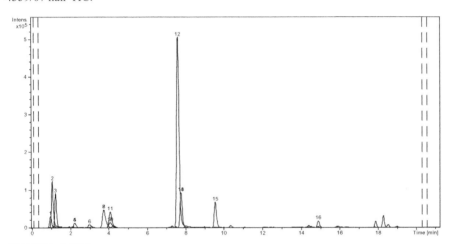

4359/07 hair EICs.

Found	Compound Name	Reg.No.	Mol.Formula	PMI	d RT [min]	Err [ppm]	Err [mDa]	Sigma	Area	Intens.	RT exp.[min]	RT meas.[min]	m/z calc.	m/z meas.
3	CAFFEINE	1451	C 8 H 10 N 4 O 2	[M+H]+	-0.05	0.3	0.1	0.002	443754	41510		4.05	4.10	195.0877	195.0876
3	THEOPHYLLINE	1454	C 7 H 8 N 4 O 2	[M+H]+	0.05	2.7	0.5	0.005	26414	3337		2.20	2.15	181.0720	181.0715
3	CODEINE	1831	C 18 H 21 N 1 O 3	[M+H]+	-0.09	0.2	-0.1	0.001	118225	12451		2.16	2.25	300.1594	300.1595
3	AMPHETAMINE	4794	C 9 H 13 N 1	[M+H]+	-0.08	4.6	-0.7	0.005	94364	8696		2.93	3.01	136.1121	136.1127
3	NICOTINE	4931	C 10 H 14 N 2	[M+H]+	-0.07	0.6	-0.1	0.002	961966	122657		1.02	1.09	163.1230	163.1231
3	COTININE	4932	C 10 H 12 N 2 O 1	[M+H]+	0.06	2.0	0.4	0.005	614524	90429		1.32	1.26	177.1022	177.1019
1	DIETHYLAMINOETHANOL	6133	C 6 H 15 N 1 O 1	[M+H]+	-1.00	6.2	-0.7	0.002	227702	29649		0.00	1.00	118.1226	118.1234
1	NORALPRENOLOL	7633	C 14 H 21 N 1 O 2	[M+H]+	-4.09	0.8	0.2	0.003	171701	13709		0.00	4.09	236.1645	236.1643
3	TRAMADOL	8861	C 16 H 25 N 1 O 2	[M+H]+	-0.04	0.8	-0.2	0.009	4669147	503776		7.56	7.60	264.1958	264.1960
2	O-DESMETHYLTRAMADOL	8862	C 15 H 23 N 1 O 2	[M+H]+	-0.13	0.9	0.2	0.003	554114	46402		3.63	3.76	250.1802	250.1799
3	NORTRAMADOL	8863	C 15 H 23 N 1 O 2	[M+H]+	-0.08	1.3	-0.3	0.005	777990	93001		7.69	7.77	250.1802	250.1805
3	O-DESMETHYLNORTRAMADOL	8864	C 14 H 21 N 1 O 2	[M+H]+	-0.13	0.8	0.2	0.002	171701	13709		3.96	4.09	236.1645	236.1643
1	DINORVENLAFAXINE	9844	C 15 H 23 N 1 O 2	[M+H]+	-3.76	0.9	0.2	0.003	554114	46402		0.00	3.76	250.1802	250.1799
1	DINORVENLAFAXINE	9844	C 15 H 23 N 1 O 2	[M+H]+	-7.77	1.3	-0.3	0.005	777990	93001		0.00	7.77	250.1802	250.1805
1	NORPROPAFENONE	12933	C 18 H 21 N 1 O 3	[M+H]+	-2.25	0.2	-0.1	0.001	118225	12451		0.00	2.25	300.1594	300.1595
3	DIBENZEPIN	99999	C 18 H 21 N 3 O 1	[M+H]+	-0.05	1.2	-0.4	0.014	561091	67192		9.50	9.55	296.1757	296.1761
1	DME	99999	C 10 H 15 N 1 O 3	[M+H]+	-14.86	2.0	-0.4	0.041	134225	17278		0.00	14.86	198.1125	198.1129

Case 4359/07 hair results report.

Figure 11.3. TIC, EICs, and results report for a hair sample analyzed by the LC/MS-TOF method (see text).

addition to the main drugs of abuse findings (codeine and amphetamine), plenty of information was obtained on the person's use of other substances. Tramadol, nicotine, and caffeine were identified together with their metabolites. The example also demonstrates the performance of the present type of data analysis, as amphetamine eluting at 3 min is readily identified under a heavy background.

11.6   LC/TOF-MS SCREENING FOR DESIGNER DRUGS IN SEIZED SAMPLES

Analysis of seizures for controlled drugs is a regular part of the duties of forensic science laboratories. These investigations often involve the use of a multitude of analytical techniques, such as spot tests, infrared spectrometry, thin-layer chromatography (TLC), GC and GC/MS, and chemometrics [28, 29]. Electron ionization GC/MS largely gains its power from the comprehensive spectral libraries that are commercially available and contain the spectra of illicit drugs and legal substances, among others. To keep in touch with the current drug scene, it is essential for laboratories to obtain the latest editions of MS libraries. This is further emphasized by the fact that reference substances for designer drugs or "new synthetic drugs" are either not generally available or their supply is limited due to administrative and time restrictions.

We have demonstrated that street drugs can be rapidly analyzed in the absence of reference standards by a combination of two modern techniques: formula-based identification by LC/TOF-MS, as shown above, and quantification by chemiluminescence nitrogen detection (LC/CLND) [30]. As the formula database for LC/TOF-MS is easily updatable with the literature data for current substances, such as designer drugs, the method is capable of responding to urgent analysis needs.

Table 11.1 shows the results of analysis by the LC/TOF-MS method on 15 seized street drug samples. In this experiment, identification was performed by accurate mass and SigmaFit™ only, without utilizing the retention times of reference standards since most of them were not available. The target database comprised 995 compounds, including the formula data from Alexander and Ann Shulgin's books *PiHKAL* [31], containing 179 phenethylamines, and *TiHKAL*, containing 55 tryptamines [32]. Sample preparation was carried out by simply dissolving the solid material in methanol (1 mg/ml) and diluting with the mobile phase 1 : 100, followed by direct analysis in a single run without further dilutions. The composition of the street drug samples, consisting of only one or a few active components and adulterants, is relatively simple compared to that of biological samples. However, the results indicate that several candidates with an identical mass are present in the target database, suggesting the need for additional analytical techniques for further resolution. The correct findings based on the confirmation analysis by TLC, GC-MS, and infrared spectrometry are marked in bold in Table 11.1. Even so, the present "dilute-and-shoot" approach by LC/TOF-MS provides a rapid and flexible characterization of the components of seized samples that is difficult to obtain by any other technique without reference standards.

TABLE 11.1. Seized Street Drugs Analyzed on a "Dilute-And-Shoot" Basis by the LC/TOF-MS Method without Reference Standards

Sample	Compound Name	Chemical Name	Structural Formula	m/z Theoretical	Mass Error [ppm]	Mass Error [mDa]	Sigma Fit™
1	DPT	N,N-dipropyltryptamine		245.2012	3.23	0.79	0.0092
	DIPT	N,N-diisopropyltryptamine		245.2012	3.23	0.79	0.0092
2	Dextromethorphan			272.2009	4.13	1.12	0.0082

(Continued)

TABLE 11.1. (*Continued*)

Sample	Compound Name	Chemical Name	Structural Formula	m/z Theoretical	Mass Error [ppm]	Mass Error [mDa]	Sigma Fit™
3	2C-E	**4-ethyl-2,5-dimethoxyphenethylamine**		**210.1489**	**2.14**	**0.45**	**0.0049**
	HMEA	*N*-ethyl-4-hydroxy-3-methoxyamphetamine		210.1489	2.14	0.45	0.0049
	DOM	4-methyl-2,5-dimethoxyamphetamine		210.1489	2.14	0.45	0.0049

			210.1489	2.14	0.45	0.0049
Ψ-DOM	2,6-dimethoxy-4-methylamphetamine					
2C-G	2,5-dimethoxy-3,4-dimethylphenethylamine		210.1489	2.14	0.45	0.0049
Methyl-DMA	2,5-dimethoxy-N-methylamphetamine		210.1489	2.14	0.45	0.0049

(Continued)

TABLE 11.1. (Continued)

Sample	Compound Name	Chemical Name	Structural Formula	m/z Theoretical	Mass Error [ppm]	Mass Error [mDa]	Sigma Fit™
4							
	5-MeO-AMT	**5-methoxy-α-methyltryptamine**		**205.1335**	**1.65**	**0.34**	**0.0056**
	4-OH-DMT	N,N-dimethyl-4-hydroxytryptamine		205.1335	1.65	0.34	0.0056
	5-OH-DMT	N,N-dimethyl-5-hydroxytryptamine		205.1335	1.65	0.34	0.0056
	5-MeO-NMT	5-methoxy-N-methyltryptamine		205.1335	1.65	0.34	0.0056

5	DPT DPT	N,N-dipropyltryptamine		**245.2012** 245.2012	**3.88** −1.75	**0.95** −0.43	**0.0125** 0.0095
6	DIPT DIPT	N,N-diisopropyltryptamine		245.2012 245.2012	3.88 −1.75	0.95 −0.43	0.0125 0.0095
	5-MeO-DIPT	**N,N-diisopropyl- 5-methoxytryptamine**		**275.2118**	**4.96**	**1.37**	**0.0101**

(Continued)

TABLE 11.1. (*Continued*)

Sample	Compound Name	Chemical Name	Structural Formula	m/z Theoretical	Mass Error [ppm]	Mass Error [mDa]	Sigma Fit™
7	NMT	N-methyltryptamine		175.1230	2.87	0.50	0.0086
8	α-MT	α-methyltryptamine		175.1230	2.87	0.50	0.0086
9	2-CI	2,5-dimethoxy-4-iodophenethylamine		308.0142	2.36	0.73	0.0109
	5-MeO-MIPT	N-isopropyl-5-methoxy-N-methyltryptamine		247.1805	5.79	1.43	0.0045

4-MeO-MIPT	N-isopropyl-4-methoxy-N-methyltryptamine	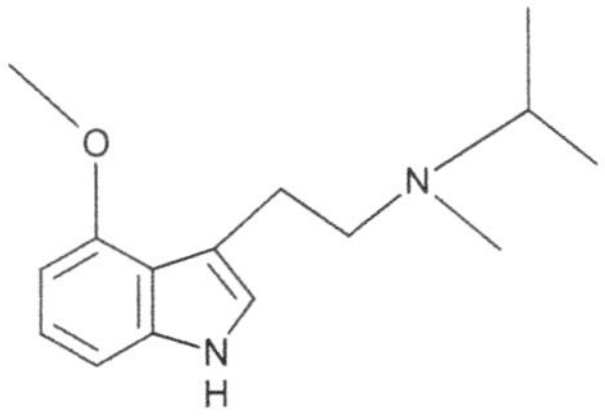 247.1805	5.79	1.43	0.0045
5-MeO-DET	N,N-diethyl-5-methoxytryptamine	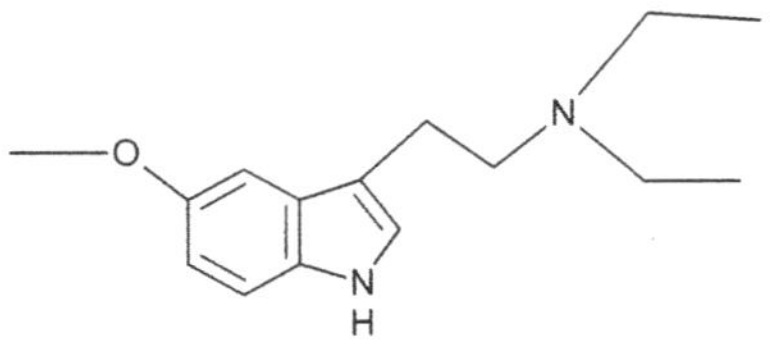 247.1805	5.79	1.43	0.0045
DPT	**N,N-dipropyltryptamine**	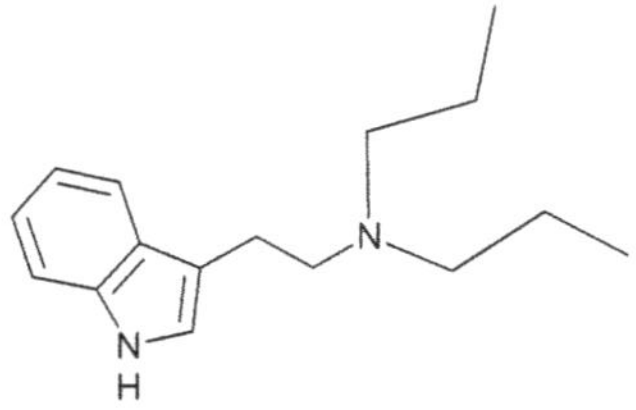 **245.2012**	**2.73**	**0.67**	**0.0099**
DIPT	N,N-diisopropyltryptamine	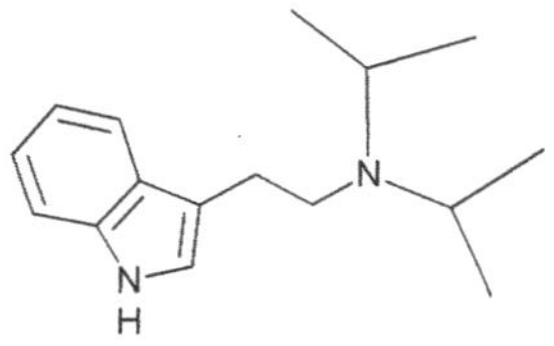 245.2012	2.73	0.67	0.0099

(Continued)

10

TABLE 11.1. (*Continued*)

Sample	Compound Name	Structural Formula	m/z Theoretical	Mass Error [ppm]	Mass Error [mDa]	Sigma Fit™
11						
	Morphine		286.1438	4.30	1.23	0.0053
	Codeine		300.1594	3.51	1.05	0.0029
	Noscapine		414.1547	0.76	0.31	0.0166

2C-T-7	**4-propylthio-2,5-dimethoxyphenethylamine**	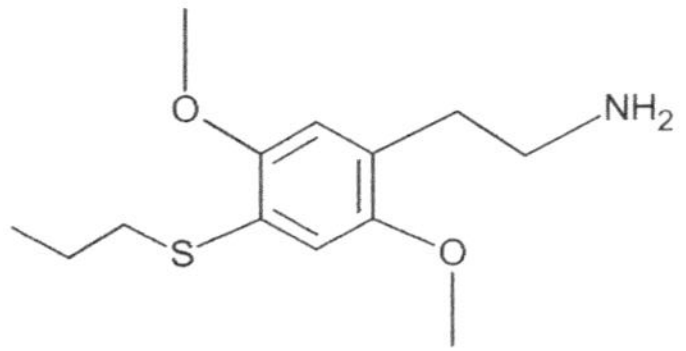	**256.1366**	**4.72**	**1.21**	**0.0120**
TP	4-propylthio-3,5-dimethoxyphenethylamine	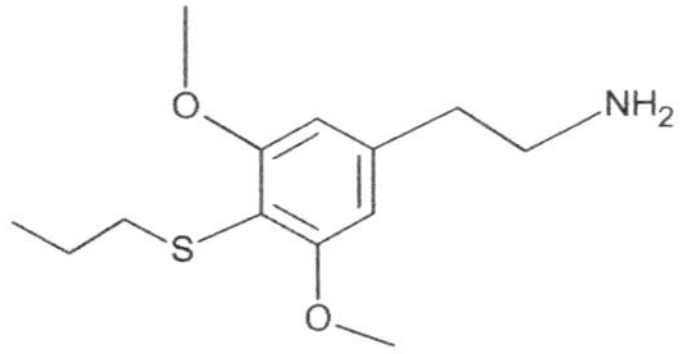	256.1366	4.72	1.21	0.0120
2C-T-4	4-isopropylthio-2,5-dimethoxyphenethylamine	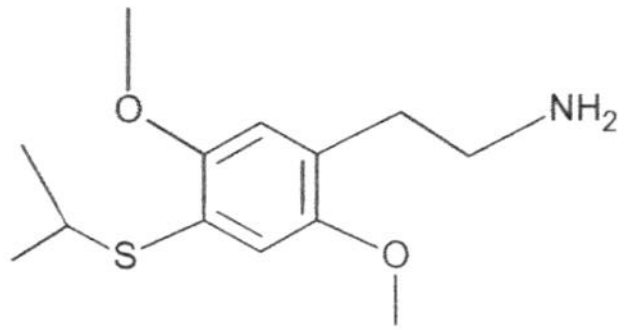	256.1366	4.72	1.21	0.0120
Ψ-2C-T-4	4-isoprylthio-2,6-dimethoxyphenethylamine	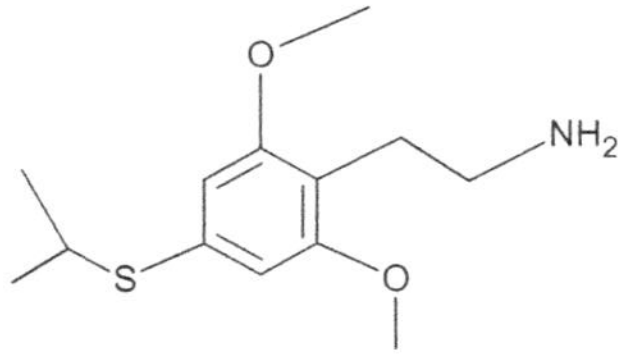	256.1366	4.72	1.21	0.0120
3-TASB	4-ethoxy-3-ethylthio-5-methoxyphenethylamine	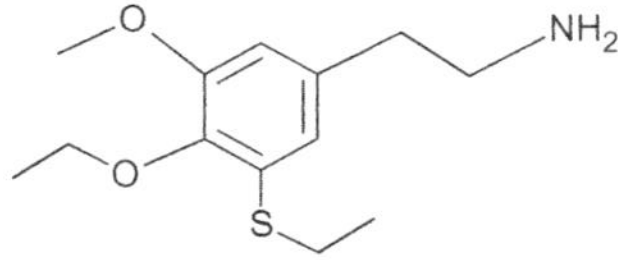	256.1366	4.72	1.21	0.0120

(Continued)

12

TABLE 11.1. (Continued)

Sample	Compound Name	Chemical Name	Structural Formula	m/z Theoretical	Mass Error [ppm]	Mass Error [mDa]	Sigma Fit™
	4-TASB	3-ethoxy-4-ethylthio-5-methoxyphenethylamine		256.1366	4.72	1.21	0.0120
	5-TASB	3,4-diethoxy-5-methylthiophenethylamine		256.1366	4.72	1.21	0.0120
	3-TSB	3-ethoxy-5-ethylthio-4-methoxyphenethylamine		256.1366	4.72	1.21	0.0120
	4-TSB	3,5-diethoxy-4-methylthiophenethylamine		256.1366	4.72	1.21	0.0120
13	**5-MeO-DMT**	**5-methoxy-N,N-dimethyltryptamine**		219.1492	1.99	0.44	0.0056

14	4-OH-MET	N-ethyl-4-hydroxy-N-methyltryptamine		219.1492	1.99	0.44	0.0056
	αNO-TMS	α,N-dimethyl-5-methoxytryptamine		219.1492	1.99	0.44	0.0056
	5-MeOPP	**p-methoxyphenylpiperazine**		**193.1335**	**5.01**	**0.97**	**0.0016**
15	MDBP	methylenedioxybenzylpiperazine		221.1295	−0.69	−0.15	0.0042
	m-CPP	**1-(3-chlorophenyl)piperazine**		**197.084**	**3.21**	**0.63**	**0.0122**

The solid material was dissolved in methanol (1 mg/mL) and diluted with the mobile phase 1:100, followed by direct analysis in a single run without further dilutions. Correct findings based on confirmation with several other methods are marked in bold. Sample 11 is opium.

11.7 CONCLUSIONS

Current LC/TOF-MS technology has generated affordable and robust bench-top instruments allowing automated accurate mass measurement in a variety of contexts, including comprehensive toxicological drug screening in urine, vitreous humor, and hair, and within forensic science. As the nominal mass accuracy of the instrument brands is similar, the remaining critical issues are the applicable dynamic range, accuracy of isotopic pattern measurement and appropriate software amenable to high-throughput screening. While retention times are utilized whenever available, accurate mass measurement can permit rapid tentative substance identification, even without reference standards. This is particularly important with new drugs, designer drugs, and their metabolites.

REFERENCES

1. Brönstrup, M., High-throughput mass spectrometry for compound characterization in drug discovery, *Top. Curr. Chem.*, **2003**, 225, 283–302.
2. Kasprzyk-Hordern, B.; Dinsdale, R.M.; Guwy, A.J., Multi-residue method for the determination of basic/neutral pharmaceuticals and illicit drugs in surface water by solid-phase extraction and ultra performance liquid chromatography-positive electrospray ionisation tandem mass spectrometry, *J. Chromatogr. A*, **2007**, 1161, 132–145.
3. García-Reyes, J.F.; Hernando, M.D.; Ferrer, C.; Molina-Díaz, A.; Fernández-Alba, A.R., Large scale pesticide multiresidue methods in food combining liquid chromatography-time-of-flight mass spectrometry and tandem mass spectrometry. *Anal. Chem.*, **2007**, 79, 7308–7323.
4. Thevis, M.; Schänzer, W., Mass spectrometry in sports drug testing: structure characterization and analytical assays. *Mass Spectrom. Rev.*, **2007**, 26, 79–107.
5. Smith, M.L.; Vorce, S.P.; Holler, J.M.; Shimomura, E.; Magluilo, J.; Jacobs, A.J.; Huestis, M.A., Modern instrumental methods in forensic toxicology. *J. Anal. Toxicol.*, **2007**, 31, 237–253.
6. Bristow, A.W., Accurate mass measurement for the determination of elemental formula—A tutorial. *Mass Spectrom. Rev.*, **2006**, 25, 99–111.
7. Stroh, J.G.; Petucci, C.J.; Brecker, S.J.; Huang, N.; Lau, J.M., Automated sub-ppm mass accuracy on an ESI-TOF for use with drug discovery compound libraries. *J. Am. Soc. Mass Spectrom.*, **2007**, 18, 1612–1616.
8. Maurer, H.H., Hyphenated mass spectrometric techniques—indispensable tools in clinical and forensic toxicology and in doping control. *J. Mass Spectrom.*, **2006**, 41, 1399–1413.
9. Baselt, R.C., *Disposition of Toxic Drugs and Chemicals in Man*, 6th ed.; Biomedical Publications: Foster City, **2002**.
10. Pajouhesh, H.; Lenz, G.R. Medicinal chemical properties of successful central nervous system drugs. *NeuroRx*, **2005**, 2, 541–553.
11. Rasanen, I.; Kontinen, I.; Nokua, J.; Ojanperä, I.; Vuori, E., Precise gas chromatography with retention time locking in comprehensive toxicological screening for drugs in blood. *J. Chromatogr. B Analyt. Technol. Biomed. Life Sci.*, **2003**, 788, 243–250.
12. Pragst, F.; Herzler, M.; Erxleben, B.T., Systematic toxicological analysis by high-performance liquid chromatography with diode array detection (HPLC-DAD). *Clin. Chem. Lab. Med.*, **2004**, 42, 1325–1340.
13. Ferrer, I.; Thurman, E.M.; Zweigenbaum, J.A., Screening and confirmation of 100 pesticides in food samples by liquid chromatography/tandem mass spectrometry. *Rapid Comm. Mass Spectrom.*, **2007**, 21, 3869–3882.
14. Maurer, H.H. Current role of liquid chromatography-mass spectrometry in clinical and forensic toxicology. *Anal. Bioanal. Chem.*, **2007**; 388: 1315–1325.

15. Gergov, M.; Weinmann, W.; Meriluoto, J.; Uusitalo, J.; Ojanperä, I., Comparison of product ion spectra obtained by liquid chromatography/triple-quadrupole mass spectrometry for library search. *Rapid Commun. Mass Spectrom.*, **2004**, 18, 1039–1046.

16. Mueller, C.A.; Weinmann, W.; Dresen, S.; Schreiber, A.; Gergov, M., Development of a multi-target screening analysis for 301 drugs using a QTrap liquid chromatography/tandem mass spectrometry system and automated library searching. *Rapid Commun. Mass Spectrom.*, **2005**, 19, 1332–1338.

17. Bristow, A.W.; Webb, K.S.; Lubben, A.T.; Halket, J., Reproducible product-ion tandem mass spectra on various liquid chromatography/mass spectrometry instruments for the development of spectral libraries. *Rapid Commun. Mass Spectrom.*, **2004**, 18, 1447–1454.

18. Maurer, H.H.; Peters, F.T., Analysis of therapeutic drugs with LC-MS; In *Applications of LC-MS in Toxicology*, Ed. Polettini A.; Pharmaceutical Press: London, **2006**, pp. 131–147.

19. Ojanperä, I.; Pelander, A.; Laks, S.; Gergov, M.; Vuori, E.; Witt, M., Application of accurate mass measurement to urine drug screening. *J. Anal. Toxicol.*, **2005**, 29, 34–40.

20. Ojanperä, S.; Pelander, A.; Pelzing, M.; Krebs, I.; Vuori, E.; Ojanperä, I., Isotopic pattern and accurate mass determination in urine drug screening by liquid chromatography/time-of-flight mass spectrometry. *Rapid Commun. Mass Spectrom.*, **2006**, 20, 1161–1167.

21. Ferrer, I.; Fernandez-Alba, A.; Zweigenbaum, J.A.; Thurman, E.M., Exact-mass library for pesticides using a molecular-feature database. *Rapid Commun. Mass Spectrom.*, **2006**, 20, 3659–3668.

22. Gergov, M.; Boucher, B.; Ojanperä, I.; Vuori, E., Toxicological screening of urine for drugs by liquid chromatography/time-of-flight mass spectrometry with automated target library search based on elemental formulas. *Rapid Commun. Mass Spectrom.*, **2001**, 15, 521–526.

23. Pelander, A.; Ojanperä, I.; Laks, S.; Rasanen, I.; Vuori, E., Toxicological screening with formula-based metabolite identification by liquid chromatography/time-of-flight mass spectrometry. *Anal. Chem.*, **2003**, 75, 5710–5718.

24. Kolmonen, M.; Leinonen, A.; Pelander, A.; Ojanperä, I., A general screening method for doping agents in human urine by solid phase extraction and liquid chromatography/time-of-flight mass spectrometry. *Anal. Chim. Acta*, **2007**, 585: 94–102.

25. Dolan, K.; Rouen, D.; Kimber, J., An overview of the use of urine, hair, sweat and saliva to detect drug use. *Drug Alcohol Rev.*, **2004**, 23, 213–217.

26. Quintela, O.; Andrenyak, D.M.; Hoggan, A.M.; Crouch, D.J., A validated method for the detection of Delta(9)-tetrahydrocannabinol and 11-nor-9-carboxy-Delta(9)-tetrahydrocannabinol in oral fluid samples by liquid chromatography coupled with quadrupole-time-of-flight mass spectrometry. *J. Anal. Toxicol.*, **2007**, 31, 157–164.

27. Gottardo, R.; Fanigliulo, A.; Bortolotti, F.; De Paoli, G.; Pascali, J.P.; Tagliaro, F., Broad-spectrum toxicological analysis of hair based on capillary zone electrophoresis-time-of-flight mass spectrometry. *J. Chromatogr. A*, **2007**, 1159, 190–197.

28. Chiarotti, M.; Fucci, N., Comparative analysis of heroin and cocaine seizures. *J. Chromatogr. B. Biomed. Sci. Appl.*, **1999**, 733, 127–136.

29. Dams, R.; Benijts, T.; Lambert, W.E.; Massart, D.L.; De Leenheer, A.P., Heroin impurity profiling: trends throughout a decade of experimenting. *Forensic Sci. Int.*, **2001**, 123, 81–88.

30. Laks, S.; Pelander, A.; Vuori, E.; Ali-Tolppa, E.; Sippola, E.; Ojanperä, I., Analysis of street drugs in seized material without primary reference standards. *Anal. Chem.*, **2004**, 76, 7375–7379.

31. Shulgin, A.; Shulgin, A., *PiHKAL. A Chemical Love Story*; Transform Press: Berkeley, **1991**.

32. Shulgin, A.; Shulgin, A., *TiHKAL. The Continuation*; Transform Press: Berkeley, **1997**.

CHAPTER 12

APPLICATION OF UPLC/TOF-MS FOR RESIDUE ANALYSIS OF VETERINARY DRUGS AND GROWTH-PROMOTING AGENTS IN PRODUCTS OF ANIMAL ORIGIN

Alida (Linda) A.M. Stolker

RIKILT-Institute of Food Safety, Wageningen, The Netherlands

THE ANALYTICAL techniques used for residue analysis of veterinary drugs and growth promoting agents are moving from target orientated methods, mainly based on liquid chromatography in combination with triple quadrupole mass spectrometric detection (LC–QqQ-MS), towards accurate mass full scan MS techniques such as time-of-flight (TOF)-MS. The full scan MS techniques enable retrospective (without re-injection of the sample) and real multi-compound (including different classes of veterinary drugs) analysis. The application of full scan MS has also consequences for the separation and extraction techniques. These have to be applicable to a broad range of compounds differing in physical and chemical properties. Experience shows that TOF-MS in combination with ultra-performance LC (UPLC) is very powerful for multi-residue analysis.

The introduction of new MS techniques has also consequences for the EU criteria defined for confirmation of the identity of compounds. It is suggested that both mass resolution and mass accuracy data are of influence for identity confirmation and both parameters should be implemented in the revision of Commission Decision 2002/657/EC.

Liquid Chromatography Time-of-Flight Mass Spectrometry: Principles, Tools, and Applications for Accurate Mass Analysis, Edited by Imma Ferrer and E. Michael Thurman
Copyright © 2009 John Wiley & Sons, Inc.

12.1   INTRODUCTION

In modern agricultural practice, veterinary drugs are being used on a large scale and administered as feed additives or via the drinking water in order to prevent the outbreak of diseases. In addition, veterinary drugs are given in the case of disease, for dehydration purposes, or to prevent losses during transportation. Growth-promoting agents such as hormones and certain veterinary drugs, mainly antibiotics, are applied to stimulate the growth of animals. The EU has strictly regulated controls on the use of veterinary drugs including growth-promoting agents, particularly in food animal species, by issuing several Regulations and Directives, and, since 1998, has prohibited antibiotics used in human medicine from being added to feed. The use of veterinary drugs is regulated through Council Regulation 2377/90/EC [1]. This regulation describes the procedure for the establishment of maximum residue limits (MRLs) for veterinary medicinal products in foodstuffs of animal origin. Its Annexes present the following information:

- Annex I includes substances with MRL values
- Annex II includes substance for which it is not considered necessary to establish MRL values
- Annex III includes substances with provisional MRL values
- Annex IV includes substances for which no MRL could be established, i.e., prohibited substances.

The prohibition of the use of growth-promoting agents such as, e.g., hormones and beta-agonists is laid down in Council Directive 96/22/EC [2]. Council Directive 96/23/EC [3] contains guidelines for controlling veterinary drug residues in animals and their products with detailed procedures for EU member states to set up national monitoring plans, including details on sampling procedures. For any type of animal or food, there are two main groups, Group A and Group B, of substances that must be monitored. Group A comprises prohibited substances (in conformity with [2] and Annex IV of [4]) and Group B comprises all registered veterinary drugs in conformity with Annexes I and III of [4] and other residues as summarized in Table 12.1. Control for Group A is more critical, i.e., has a higher priority, because of public-health concern: relatively large numbers of samples have to be analyzed and more stringent criteria have to be used [5] in view of the serious implications of positive results for public health.

Technical guidelines and performance criteria, e.g., detection level, selectivity, and specificity, for residue control in the framework of Directive 96/23/EC are described in Commission Decision 2002/657/EC [5]. Next to the general performance requirements, additional requirements are described for confirmatory methods by introducing the concept of identification points (IPs) and defining criteria for ion intensities. During confirmatory analysis a specific number of IPs has to be collected. For the confirmation of the identity of Group A substances—commonly referred to as unauthorized, illegal, banned or prohibited substances—a minimum of four IPs is required. For the confirmation of the identity of substances listed in Group B, a minimum of three IPs is required. The number of IPs earned by a specific analysis depends on the technique used. However, almost invariably these techniques have

TABLE 12.1. Group A and B Substances

Group

Group A: substances having anabolic effects and unauthorized substances

- Stilbenes, stilbene derivatives, and their salts and esters
- Antithyroid agents
- Steroids
- Resorcylic acid lactones including zeranol
- β-Agonists
- Compounds included in Annex IV to Council Regulation 2377/90/EC [1]

Group B: veterinary drugs and contaminants

- Antibacterial substances, including sulfonamides, quinolones
- Other veterinary drugs
  - Anthelmintics
  - Anticoccidiostats, including nitroimidazoles
  - Carbamates and pyrethroids
  - Carbadox and olaquindox
  - Sedatives
  - Non-steroidal anti-inflammatory drugs (NSAIDs)
  - Other pharmacologically active substances
- Other substances and environmental contaminants
  - Organochlorine compounds including PCBs
  - Organophosphorus compounds
  - Chemical elements
  - Mycotoxins
  - Dyes
  - Others

to be based on mass spectrometric detection. A low-resolution mass spectrometer, such as a triple quadrupole (QqQ) or an ion-trap (IT), is able to acquire 1.0 IP for the precursor ion and 1.5 IPs for each product ion; that is, with the selection of two multi-reaction-monitoring (MRM) transitions, 4.0 IPs are acquired. In case the whole MS-MS spectra is monitored, the IPs acquired depend on the number, n, of fragment ions present on the product-ion scan (PIS) spectrum [i.e., (1 + 1.5n)]. The mass accuracy of a high-resolution mass spectrometer (HRMS) acquires 2.0 IPs for the precursor ion and 2.5 for each product ion, so, in this case, the IPs acquired in the PIS mode are (2 + 2.5n).

Within [5] some new definitions like MRPL, CCα and CCβ are introduced, these definitions will be shortly discussed here. The minimum required performance limit (MRPL) describes the concentration level at which a non-authorized substance has to be measured. For residues of non-authorized substances, the legal tolerance is zero. This level, however, can not be measured and within decision 2002/657/EC this level is defined by the MRPL.

Decision limit, CCα, and detection capability, CCβ, are intended to replace the following method characteristics: limit of detection (LOD) and limit of quantification (LOQ). CCα is defined as "the concentration at and above which it can be concluded with an error probability of α that a sample is non-compliant (positive)." CCβ is defined as "the smallest content of the substance that may be detected, identified and/or quantified in a sample with an error probability of β." In β% of the cases, a non-compliant sample will be classified as compliant, and therefore reveals a false-negative result.

In 2005 an extended review was published [7] describing the analytical strategies for residue analysis of veterinary drugs and growth-promoting agents in food-producing animals. Since then, more than 70 papers on veterinary drug residue analysis have been published [6]. An emerging trend is recognized in the residue analysis from liquid chromatography in combination with triple quadrupole mass spectrometric detection (LC–QqQ-MS) based screening and confirmation towards accurate mass full scan MS techniques like time-of-flight (TOF)-MS [8]. The accurate-mass capability allows the reconstruction of highly selective accurate-mass chromatograms of target residues in complex matrices. This is of special interest for confirmation analysis and identification of "unknowns." The full scan MS technologies offer also the advantage of retrospective analysis without re-analysis. Since the introduction of full (accurate-)mass MS techniques a clear trend towards multi-residue methods can be observed. Full mass MS makes it possible that hundreds of different compounds can be screened in a single analysis. For such comprehensive analyses not only the detection but also the extraction and separation have to be optimized. A very generic sample extraction method is necessary to extract a broad range of compounds with different physical/chemical properties. Next, for efficient separation, more advanced techniques like ultra performance liquid chromatography (UPLC) have to be introduced.

This chapter will give an overview of some interesting applications published recently regarding the use of (UP)LC–TOF-MS for the analysis of residues of veterinary drugs and growth promoting agents. The overview is not exhaustive, however, from our own experience and from the information provided in the literature one may conclude that the examples given are representative for the new analytical trends in this working field.

Finally a discussion is initiated regarding the number of IPs acquired by using new analytical techniques like LC–TOF-MS.

12.2 MULTI-COMPOUND METHODS

12.2.1 QqQ- Versus TOF-MS

The cost-effectiveness of analytical procedures is becoming an important issue for all laboratories involved in residue analysis. Automation has been introduced to speed up many analytical procedures but instrumentation is expensive, and there should be a distinct need to test large numbers of samples to justify such significant capital expenditure. An alternative way to improve cost-effectiveness is to maximize the number of analytes that may be determined by a single procedure or from a single portion of test material. Such an approach is extremely effective when multi-residue techniques, such as LC–MS/MS are used. However, most reported multi-residue methods target a few closely related compounds, usually belonging to a single drug class [9–14]. However, there are few procedures which describe methods that can analyze compounds from unrelated classes of drugs. Granelli et al. [15] reported a multi-residue screening method, at MRL level, for antibiotics in meat and kidney. The method includes 19 antibiotics from five classes, i.e., tetracyclines, sulphonamides, quinolones, β-lactams, and macrolides. After LLE with 70% methanol, the extracts are analyzed by LC–MS/MS. Another multi-residue method was described by Yamada et al. [16]. The authors developed a simple and rapid method using LC–MS/MS for the simultaneous determination of 130 veterinary drugs and their metabolites in bovine, porcine, and chicken muscle. The drugs (1–10 µg/kg muscle) were extracted with acetonitrile–methanol (95 : 5, v/v). The extracts were defatted by adding n-hexane saturated with acetonitrile and, next, evaporated, dissolved in methanol, analyzed by LC with gradient elution on a C18-bonded phase, and detected by QqQ-MS. The LODs ranged from 0.03 to 3 µg/kg and the quantification limits from 0.1 to 10 µg/kg. Over 100 drugs (sulphonamides, beta-lactams, quinolones, coccidiostatics, steroids, etc.) could be analyzed in muscle tissue.

The key aspects in multi-residue methods include sensitive and selective detection. Modern triple QqQ-MS instrumentation permits excellent sensitivity and selectivity. Modern instruments produce high signal to noise ratio, even when relying on short MRM dwell times. This permits the simultaneous monitoring of an increasing number of transitions [17]. However, increasing the number of analyte peaks to be monitored beyond 50–80 requires multiple injections or monitoring specific transitions at a specific retention time window [18]. Methods based on retention time windows, require frequent readjustments due to small shifts of retention times. An attractive alternative is the use of full mass scan MS technique, for example by using TOF, Fourier-transform ion cyclotron resonance (FTICR) or Fourier-transform Orbitrap (FT Orbitrap) MS. The medium to high resolution—for example for TOF-MS—of 10,000 FWHM affects significantly the selectivity and therefore the sensitivity gain, compared to unit-resolution scanning MS instrumentation. A significant advantage of TOF-MS is that no *a priori* hypothesis about the presence of certain drugs is required; that is, no analyte-specific transitions have to be defined before injecting the sample. The high-resolution, full scan data permit the testing of any *a posteriori* hypotheses by extracting any desired exact mass chromatogram. Moreover, the accurate mass capability of LC–TOF-MS allows the reconstruction of highly selective accurate mass chromatograms of target residues in complex matrices. Recently an interesting publication by Hernando et al. [20] demonstrated the use of LC–TOF-MS for the simultaneous determination of the residues of antibiotics (enrofloxacin, oxolinic acid, flumequine, erythromycin), fungicides (MG and LMG) and a parasiticide (emamectin benzoate) in the edible portion of salmon. TOF-MS with a mass resolution of 10,000 full peak width at half-maximum (FWHM) and an accuracy of <5 ppm was used for confirmatory purposes. LC was carried out on a C18 column. The extraction was by means of solid-liquid extraction (addition of sorbent, Bondesil-NH_2, to the liquid). For the MRL substances—with MRLs of 100–200 µg/kg—the CC_α and CC_β values were in the range of 103–217 and 107–234 µg/kg, respectively. For the non-authorized compounds MG and LMG, the CC_α and CC_β values were 8–38 and 13–65 µg/kg, respectively. Although these levels exceed the MRPL set by the EU, which is 2 µg/kg for the sum of MG and LMG. The use of LC–TOF-MS is very interesting for multi-residue screening and confirmatory analysis because "unknowns" can be detected and retrospective analysis is possible without re-analysis of the sample.

A very interesting example of the use of full scan mass MS for multi-residue screening in urine is described by Kaufmann et al. [17]. The method covers more than 100 analytes belonging to different classes of veterinary drugs, e.g., quinolones, sulphonamides, nitroimidazoles, penicillins, macrolides, benzimidazoles, tetracyclines, and tranquillizers. After sample pre-treatment by means of dilution, the samples were injected—without prior filtration—in an UPLC–TOF-MS system. The quantification limits were <10 µg/L for over 90% of the 108 tested veterinary drugs and metabolites. The highest quantification levels were found for sulfacetamide and ronidazol: both had an LOQ of 45 µg/L. The combination of UPLC and TOF provided sufficient sensitivity and selectivity. The authors concluded that the possibility of retrospective analysis is extremely valuable, especially when metabolites are monitored which are not commercially available. The detection of a parent drug and the corresponding metabolite with high mass accuracy and resolution, as well as the isotopic ratio, can provide selectivities beyond MRM ratios. The monitoring of drug-class-specific MS fragments (after CID) permits the detection of non-target analytes, which is not possible with unit-resolution MS techniques.

Accurate mass determination and calculated elemental composition data can be used for structure elucidation as well. For example, Nielen et al. [19] discussed mass resolution and accuracy for the LC–MS screening and confirmation of targeted analytes and for the identification of unknowns using an anabolic steroid, stanozolol, and a designer β-agonist, clenbuterol-R, as model substances. It was shown theoretically and experimentally that mass accuracy criteria without proper mass resolution criteria yield false compliant (false negative) results, both in the MS screening and MS/MS confirmation of stanozolol. On the other hand previous medium resolution accurate mass TOF-MS/MS data of the designer β-agonist were fully confirmed by high-resolution FT Orbitrap-MS^n experiments.

12.2.2   Separation and Extraction

In the multi-residue method using the full scan MS, the LC part of the system sometimes causes certain limitations. Due to the high number of compounds to be separated the run times will become relatively long. LC runs can be shortened by utilizing short columns and/or fast gradients. The price to be paid is low(er) chromatographic resolution, which necessitates a further reduction of the already limiting dwell times if MRM transitions are to be monitored. The introduction of pressure-stable 1.7 µm particulate packing materials and novel low-dead- volume, high-pressure (1000 bar) LC equipment provides strategies to improve resolution while maintaining or even shortening run times. This technique is called ultra-performance LC (UPLC). An essential aspect of the UPLC concept is the use of sub-2 µm particulate packing materials, while maintaining other aspects of the column geometry, e.g., column length. UPLC–TOF-MS provides significant advantages concerning selectivity, sensitivity and speed. The higher resolution provided by UPLC is an important factor to compensate for the fact that the selectivity of currently available TOF-MS instrumentation is still less than that provided by monitoring MS→MS transitions. Applications of the technique were described above [17, 19].

The use of LLE and/or SPE for sample pre-treatment is used in more than 80% of all published methods for the determination of veterinary drug residues. A specific combination of LLE and SPE can be very selective for a specific class of veterinary drugs. For real multi-residue analysis new sample pre-treatment approaches are necessary to extract drugs with different physical/chemical properties. Kaufmann et al. [17] limited themselves to dilution of the urine samples prior to UPLC–TOF-MS analysis. This approach is not applicable when very low levels of residues have to be determined, for example for the unauthorized substances, or when semi-solid or solid samples have to be analyzed. Hernando et al. [20] described a specific liquid-solid-extraction procedure. Samples of salmon were extracted with acetonitrile (plus 0.1% acetic acid). The sorbent (Bondesil-NH$_2$) was added to the primary extract to adsorb matrix components; after a centrifugation step the supernatant was analyzed by LC–TOF-MS for residues of veterinary drugs.

A multi-residue cation-exchange clean-up procedure for basic drugs is described by Stubbings et al. [21]. By using the cation-exchange Bond Elute SCX as sorbent they could isolate and concentrate sulphonamides, benzimidazoles, levamisole, nitroimidazoles, tranquillizers, and FQs from acetonitrile extracts. The extracts were obtained from matrices like animal tissues and eggs. The recoveries were in the range of 56–104% for all analyte/matrix combinations. Except for carazolol and ciprofloxacin the procedure yields recoveries which are comparable to those obtained using more selective procedures. The authors reported that the SCX clean-up can be used also for other drugs, e.g., quinoxaline carboxylic acid, dapsone, oxolinic acid, nalidixic acid and β-agonists.

The application of matrix-solid-phase-dispersion (MSPD) as a generic extraction technique was recently reviewed by Bogialli et al. [22]. The authors concluded that the technique is valuable for a wide variety of analyte/matrix combinations but that improvements, e.g., miniaturization and direct coupling with other techniques (LC or pressurized liquid extraction), are necessary to make the technique more attractive and useful for routine analysis.

12.3 THE USE OF UPLC–TOF-MS FOR MULTI-COMPOUND ANALYSIS OF VETERINARY DRUGS

This section describes and discusses the analytical method developed and validated for the screening and quantification of more than a hundred residues of veterinary drugs in milk. Furthermore, the results of the analysis of 100 milk samples will be described. For this comprehensive analysis in a complex matrix UPLC–ToF-MS was used [23]. The veterinary drugs represent different classes including benzimidazoles, macrolides, penicillins, quinolones, sulfonamides, pyrimidines,

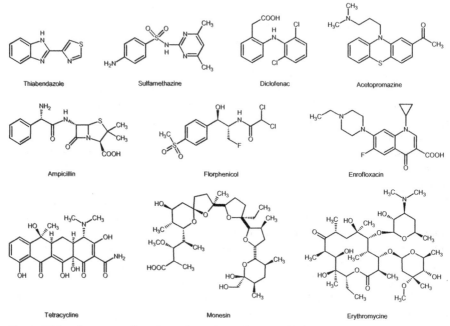

Figure 12.1. Structures of a subset of veterinary drugs studied.

tetracylines, nitroimidazoles, tranquillizers, ionophores, amphenicols, and non-steroidal anti-inflammatory agents (NSAIDs) (see Figure 12.1).

The benzimidazoles belong to the group of anthelmintic drugs acting primarily against lungworms and liver fluke. The penicillins, macrolides, quinolones, sulfonamides, tetracyclines, and amphenicols are different classes of antibiotics. The nitroimidazoles and ionophores are frequently used as feed additives for the treatment and prevention of certain bacterial and protozoal diseases. Finally, the tranquillizers and NSAIDs are included in the method. Milk was selected because it is an important matrix for residue control which can contain residues from different classes of veterinary drugs.

The details of the developed method are described in [23]. Briefly, the method is as follows: milk (2 mL) was mixed with acetonitrile (2 mL) to cause precipitation of proteins. After an intensive shaking period of 30 min, the samples were centrifuged for 15 min (3600 g at T = 10 °C). From the supernatant 2 mL was diluted 10 times with water. The diluted sample was applied to a StrataX-SPE column. The column was conditioned with 3 mL of methanol and 3 mL of water and washed with 3 ml of water. The analytes were eluted with 3 mL of methanol.

After evaporation of the StrataX eluate under a stream of nitrogen at 40 °C, the dried extract was redissolved in 50 μL acetonitrile and vortexed for 30 sec. Next 450 μL of water–formic acid (1000:2, v/v) was added and an aliquot of 40 μL was analyzed by UPLC–TOF-MS in the full scan mode. After acquisition the specific [M + H]+ ions (see Table 12.2) were extracted from the spectra. The used extraction window width was drug specific and ranged from 10 to 200 mDa.

TABLE 12.2.   Accurate Mass Measurements, Retention Times (Rt), and Validation Levels of the Studied 101 Veterinary Drugs in Milk

Compound	Element Composition	Rt(min)	[M+H]+ (m/z)	Measured (m/z)	Δppm	MRL/MRPL	VL (μg/L)
Benzimidazoles							
Albendazole	C12H15N3O2S	4.72	266.0963	266.0963	0.0	100	50
Albendazole sulfoxide	C12H15N3O3S	3.23	282.0912	282.0925	4.6	100	50
Albendazole sulfone	C12H15N3O4S	3.78	298.0862	298.0851	-3.7	100	50
Albendazole amino sulfone	C10H13N3O2S	2.46	240.0807	240.0800	-2.9	100	50
Fenbendazole	C15H13N3O2S	5.64	300.0806	300.0842	11.7	10	10
Oxfendazole	C15H13N3O3S	3.94	316.0756	316.0771	4.7	10	10
Oxfendazole sulfone	C15H13N3O4S	4.65	332.0705	332.0723	5.4	10	10
Mebendazole	C16H13N3O3	4.71	296.1035	296.1094	19.9	2	2
Amino mebendazole	C14H11N3O	3.6	238.0980	238.1002	9.2	2	2
Hydroxy mebendazole	C16H15N3O3	3.61	298.1192	298.1199	2.3	2	2
Levamisole	C11H12N2S	2.31	205.0799	205.0804	2.4	2	2
Thiabendazole	C10H7N3S	2.48	202.0439	202.0414	-12.4	100	50
5-Hydroxy thiabendazole	C10H7N3OS	2.32	218.0388	218.0378	-4.6	100	50
Flubendazole	C16H12FN3O3	4.98	314.0941	314.0962	6.7	2	2
Amino flubendazole	C14H10FN3O	3.8	256.0886	256.0912	10.2	2	2
Hydroxy flubendazole	C16H14FN3O3	3.84	316.1097	316.1083	-4.4	2	2
Oxibendazole	C12H15N3O3	3.9	250.1191	250.1187	-2.0	2	2
Macrolides							
Erythromycin	C37H67NO13	5.06	734.4690	734.4697	0.8	40	40
Spiramycin	C43H74N2O14	4.08	843.5218)	843.5172	-5.5	200	50
Tilmicosin	C46H81N2O13	4.68	869.5738	869.5701	-4.4	50	50
Tylosin	C46H77NO17	5.40	916.5270	916.5223	-5.1	50	50
Josamycin	C42H70NO15	6.14	828.4745	828.4775	3.6		50
Tiamulin	C28H47NO4S	5.45	494.3304	494.3325	4.2		50
Valnemulin	C31H52N2O5S	6.01	565.3675	565.3756	14.3		50
Lincomycin	C18H34N2O6S	2.59	407.2216	407.2227	2.7	150	50
Pirlimycin	C17H31ClN2O5S	3.93	411.1720	411.1725	1.2	100	50
Tulatromycin	C41H79N3O12	3.4	806.5742	806.5682	-7.4		50

(Continued)

TABLE 12.2. (*Continued*)

Compound	Element Composition	Rt(*min*)	[M+H]+ (*m/z*)	Measured (*m/z*)	Δppm	MRL/MRPL	VL (μg/L)
Penicillins							
Penicillin V	C16H18N2O5S	5.38	351.1014	351.1083	19.4		25
Ampicillin	C16H19N3O4S	2.79	350.1174	350.1222	13.4	4	25
Oxacillin	C19H19N3O5S	5.67	402.1123	402.1125	0.2	30	30
Nafcillin	C21H23N2O5S	6.29	415.1327	415.1322	−1.4	30	30
Cloxacillin	C19H18ClN3O5S	6.05	436.0734	436.0841	24.5	30	30
Dicloxacillin	C19H18Cl2N3O5S	6.53	470.0330	470.0370	5.5	30	30
Quinolones							
Nalidixic acid	C12H12N2O3	4.71	233.0926	233.0944	7.7		2
Oxolinic acid	C13H11NO5	3.99	262.0715	262.0731	6.1		2
Flumequine	C14H12FNO3	4.99	262.0879	262.0869	−3.8	50	50
Norfloxacin	C16H19FN3O3	2.92	320.1410	320.1420	3.1		25
Ciprofloxacin	C17H18FN3O3	3	332.1410	332.1432	6.6		5
Lomefloxacin	C17H20F2N3O3	3.11	352.1472	352.1487	4.0		5
Enrofloxacin	C19H22FN3O3	3.22	360.1723	360.1706	−4.7	100	50
Marbofloxacin	C17H20FN4O4	2.79	363.1468	363.1472	0.8	75	50
Difloxacin	C21H20F2N3O3	3.46	400.1472	400.1488	3.7		2
Danofloxacin	C19H21FN3O3	3.15	358.1567	358.1585	5.0	30	30
Sarafloxacin	C20H17F2N3O3	3.42	386.1316	386.1342	6.7		5

Sulfonamides							
Dapsone	C12H12N2O2S	3.01	249.0697	249.0703	2.0	2	5
Sulfadiazine	C10H10N4O2S	2.02	251.0602	251.0607	1.6	100	50
Sulfamethoxazole	C10H11N3O3S	3.42	254.0599	254.0604	2.0	100	50
Sulfamethazine	C12H14N4O2S	2.85	279.0915	279.0903	-4.7	100	50
Sulfadimethoxine,	C12H14N4O4S	3.53	311.0814	311.0836	7.1	100	50
Sulfadoxine	C12H14N4O4S	4.24	311.0814	311.0811	-1.0	100	50
Sulfaquinoxaline	C14H12N4O2S	4.32	301.0759	301.0752	-2.3	100	50
Sulfachloropyridazine	C10H9ClN4O2S	3.22	285.0213	285.0237	8.4	100	50
Sulfamerazine	C11H12N4O2S	2.47	265.0759	265.0762	1.1	100	50
Sulfamethizole	C9H10N4O2S2	2.89	271.0323	271.0323	0.0	100	50
Sulfamethoxypyridazine	C11H12N4O3S	2.96	281.0708	281.0711	1.1	100	50
Sulfamonomethoxine	C11H12N4O3S	3.22	281.0708	281.0710	0.7	100	50
Sulfamoxole	C11H13N3O3S	3.69	268.0756	268.0771	5.6	100	50
Sulfapyridine	C11H11N3O2S	2.38	250.0650	250.0662	4.8	100	50
Sulfisoxazole	C11H13N3O3S	2.84	268.0756	268.0762	2.2	100	50
Sulfathiazole	C9H9N3O2S2	2.31	256.0214	256.0223	3.5	100	50
Pyrimidine							
Trimethoprim	C14H18N4O3	2.77	291.1457	291.1490	3.3	50	25
Tetracyclines							
Tetracycline	C22H24N2O8	3.04	445.1611	445.1624	2.9	100	100
Doxycycline	C22H24N2O8	4.05	445.1611	445.1628	3.8	100	100
Oxytetracycline	C22H24N2O9	2.82	461.1560	461.1555	-1.1	100	100
Chlorotetracycline	C22H23ClN2O8	3.77	479.1221	479.1218	-0.6	100	100

(Continued)

TABLE 12.2. *(Continued)*

Compound	Element Composition	Rt(*min*)	[M+H]+ (*m/z*)	Measured (*m/z*)	Δppm	MRL/MRPL	VL (μg/L)
Nitroimidazolen							
Ipronidazol	C7H11N3O2	3.31	170.0930	170.0931	0.6	1	5
Hydroxy-ipronidazol	C7H11N3O3	2.85	186.0879	186.0886	3.8	1	5
Tranquillizers							
Azaperol	C19H24FN3O	2.83	330.1982	330.1982	0.0	1	1
Azaperon	C19H22FN3O	3.11	328.1825	328.1839	4.3	1	1
Propionylpromazine	C20H24N2OS	5.59	341.1688	341.1736	14.1	1	1
Acetopromazine	C19H22N2OS	5.06	327.1531	327.1549	5.5	1	1
Xylazine	C12H16N2S	3.22	221.1112	221.1109	-1.4	1	1
Haloperidol	C21H23ClFNO2	4.92	376.1480	376.1503	6.1	1	1
Chloropromazine	C17H19ClN2S	5.75	319.1036	319.1050	4.4	1	2
Carazolol	C18H22N2O2	3.94	299.1760	299.1774	4.7	1	1
NSAIDs							
Piroxycam	C15H13N3O4S	4.92	332.0705	332.0716	3.3	50	50
Propyphenazone	C14H18N2O	5.09	231.1497	231.1480	-7.4	20	20
Indoprofen	C17H15NO3	5.52	282.1130	282.1168	13.5	20	20
Tolmetin	C15H15NO3	5.93	258.1130	258.1196	25.6	50	50
Ketoprofen	C16H14O3	6.09	255.1021	255.1104	32.5	50	50
Naproxen	C14H14O3	6.12	231.1021	231.1045	10.4	50	50
Fenbufen	C16H14O3	6.39	255.1021	255.1056	13.7	50	50
Carprofen	C15H12ClNO2	6.75	274.0635	274.0670	12.8	50	50
Diclofenac	C14H11Cl2NO2	6.82	296.0250	296.0277	10.8	50	50

Niflumic acid	C13H9F3N2O2	6.66	283.0694	283.0729	12.4	50	50
Phenylbutazone	C19H20N2O2	6.91	309.1603	309.1624	6.8	25	25
Flufenamic acid	C14H10F3NO2	7.02	282.0742	282.0756	5.0	50	50
Mefenamic acid	C15H15NO2	7.03	242.1181	242.1161	−8.3	50	50
Meclofenamic acid	C14H11Cl2NO2	7.04	296.0250	296.0280	11.8	50	50
Isopyrin	C14H19N3O	2.59	246.1606	246.1606	0.0		50
Isoxicam	C14H13N3O5S	6.34	336.0654	336.0675	6.2		50
Tenoxicam	C13H11N3O4S2	3.91	338.0269	338.0287	5.3		50
Sulindac	C20H17FO3S	5.78	357.0961	357.0980	5.3		50
Indomethacin	C19H16ClNO4	6.85	358.0846	358.0857	3.1		50
Flunixin	C14H11F3N2O2	5.65	297.0851	297.0869	6.1	50	50
Ionophores							
Monensin Na	C36H61O11Na	7.84	693.4190	693.4164	−3.7		1
Salinomycine Na	C42H69O11Na	7.91	773.4816	773.4732	−10.9		1
Narasin Na	C43H72O11Na	8.05	787.4972	787.4913	−7.5		1
Maduramycine Na	C47H79O17NH4	8.07	934.5739	934.5606	−14.2		1
Amphenicols							
Florphenicol NH4	C12H14Cl2FNO4S	3.52	358.0083	358.0138	15.4		5
Thiamphenicol NH4	C12H15Cl2NO5S	2.76	356.0126	356.0125	−0.3	50	50

VL = validation level (see also text).
Adapted from [23].

For quantification, a detector response versus concentration plot was constructed. To this end, six blank milk samples were fortified with different concentrations of the specific drug. The real milk samples were analyzed together with the calibration samples (Matrix Matched Standards or MMS); concentrations were calculated using the linear regression method.

For two classes of drugs an internal standard is used. Cinchophen is the internal standard for the analysis of quinolones and nigericine for ionofores. The internal standards are added just before the protein precipitation step and the final results are corrected for losses of the internal standard.

12.3.1 Method Validation

The developed method was validated based on the procedure described in [5] for a quantitative screening method. The following characteristics were determined: repeatability, within-lab reproducibility, accuracy, linearity, and CCβ.

Repeatability, Within-Lab Reproducibility, Accuracy The validation study for the quantitative screening of veterinary drugs in milk was carried out at three concentration levels *viz.* 0.5–1.0–1.5 times "validation level". In this chapter "validation level" (VL) is defined as the MRL or MRPL concentration level. For drugs without an MRL or MRPL a specific level of interest was defined based on the drug characteristics (class of compounds) or based on MRL or MRPL of the specific drug in other matrices like liver, kidney, or meat. All concentration levels used for the validation study are described in Table 12.3. For some compounds the VL was at a slightly higher level than the MRL or MRPL level due to the limited sensitivity of the TOF detector for that specific drug. To prevent overloading of the column some compounds are validated at a concentration level below the MRL or MRPL. Blank samples of milk were fortified with a specific concentration of drug and six replicates of each sample were analyzed on one day. The procedure was repeated on two additional days.

From the data obtained within-day repeatability, within-lab reproducibility and accuracy were calculated. The accuracy—(detected concentration minus added concentration) * 100%—preferably has to be within the 70–120% range. This range is set for this study as an acceptable accuracy for a multi-compound quantitative screening method for the concentration range 1–150 µg/L.

Figure 12.2 presents, respectively, (a) the summarized repeatability, (b) reproducibility, and (c) accuracy results. Figure 12.2a and 12.2b show the %RSD (repeatability or within-lab reproducibility) versus the percentage of compounds for which the specific %RSD is obtained. Figure 12.2c presents the percentage accuracy versus the percentage of compounds showing that specific accuracy.

From Figure 12.2a it is concluded that at the validation level (VL) more than 60% of the compounds tested show a repeatability of RSD < 10% and less than 15% of the compounds show a repeatability of RSD > 20%. At 0.5 * VL the percentages are significant different. At 0.5 * VL almost 50% of the compounds shows repeatability results of RSD > 10%. Comparing the 1 * VL and 1.5 * VL levels only

TABLE 12.3. Repeatability, Within-Lab Reproducibility, Accuracy, Linearity, CCbeta, and LOQ Results of the Studied 101 Veterinary Drugs in Milk

Compound	VL (μg/L)	%RSD Within-Lab Reproducibility			Accuracy (%)				(r^2)	CCβ	LOQ
		0.5VL	1VL	1.5VL	0.5VL	1VL	1.5VL				
Benzimidazoles											
Albendazole	50	10.7	9.1	8.7	88	99	104		0.998	64.9	6.3
Albendazole sulfoxide	50	10.7	7.8	5.3	92	90	101		1	61.5	6.3
Albendazole sulfone	50	4.4	6.6	4.0	101	99	105		0.999	60.8	6.3
Albendazole amino sulfone	50	13.1	5.9	5.8	97	113	108		0.995	61.0	6.3
Fenbendazole	10	6.9	7.1	12.6	110	112	107		0.993	12.6	1.3
Oxfendazole	10	8.9	5.3	4.2	83	108	109		0.994	11.9	1.3
Oxfendazole sulfone	10	13.1	7.3	5.5	85	106	109		0.998	12.5	1.3
Mebendazole	2	19.0	9.4	8.0	90	108	107		0.999	2.7	0.3
Amino mebendazole	2	13.5	9.5	9.3	87	100	101		0.998	2.6	0.3
Hydroxy mebendazole	2	16.0	7.7	9.2	87	100	98		0.997	2.5	0.3
Levamisole	2	26.8	19.9	11.9	64	133	140		0.952	3.7	0.3
Thiabendazole	50	23.1	8.8	10.8	137	168	148		0.929	74.2	6.3
5-Hydroxy thiabendazole	50	26.4	11.4	14.2	85	139	134		0.971	76.1	6.3
Flubendazole	2	10.7	8.1	7.7	87	102	102		1	2.5	0.3
Amino flubendazole	2	14.8	8.5	4.6	89	105	105		0.999	2.6	0.3
Hydroxy flubendazole	2	17.0	11.2	10.7	86	99	99		0.998	2.7	0.3
Oxibendazole	2	8.4	6.7	5.7	85	110	112		0.99	2.5	0.3
Macrolides											
Erythromycin	40	9.2	8.6	13.1	92	100	101		0.999	51.3	5.0
Spiramycin	50	7.7	7.6	8.4	104	116	113		0.999	64.4	6.3
Tilmicosin	50	13.2	5.7	5.8	95	104	99		0.999	59.7	6.3
Tylosin	50	7.3	7.3	10.5	101	113	110		1	63.5	6.3

(Continued)

TABLE 12.3. (*Continued*)

Compound	VL (μg/L)	%RSD Within-Lab Reproducibility			Accuracy (%)			(r²)	CCβ	LOQ
		0.5VL	1VL	1.5VL	0.5VL	1VL	1.5VL			
Josamycin	50	9.5	9.2	10.6	113	109	105	0.999	66.4	6.3
Tiamulin	50	5.6	6.4	7.1	100	112	110	0.999	61.8	6.3
Valnemulin	50	6.5	9.9	13.5	127	112	105	0.997	68.2	6.3
Lincomycine	50	37.0*	32.5	24.4	72	157	155	0.872	133.6	6.3
Pirlimycine	50	6.8	18.7	10.7	105	104	104	0.99	81.9	6.3
Tulatromycine	50	23.0*	19.7	25.5	82	106	105	0.99	84.2	6.3
Penicillins										
Penicillin V	25	15.1	13.3	7.6	104	108	143	0.999	67.8	3.1
Ampicillin	25	35.2	33.8	22.2	98	124	116	0.991	59.3	3.1
Oxacillin	30	10.5	10.9	13.3	119	108	100	0.999	41.6	3.8
Nafcillin	30	39.6	34.8	33.1	71	76	84	1	56.0	3.8
Cloxacillin	30	31.7	25.0	17.5	113	112	108	1	57.5	3.8
Dicloxacillin	30	13.0	8.6	15.1	97	105	101	0.999	38.9	3.8
Quinolones										
Nalidixic acid	2	16.7	22.5	20.3	96	99	99	0.993	3.5	0.3
Oxolinic acid	2	6.9	20.0	18.0	98	99	98	0.996	3.3	0.3
Flumequine	50	12.1	16.2	12.3	115	92	84	0.998	74.5	6.3
Norfloxacin	25	20.0	28.9	27.1	109	114	114	0.986	52.0	3.1
Ciprofloxacin	5	16.6	27.5	28.7	100	99	100	0.995	9.5	0.6
Lomefloxacin	5	15.9	26.4	24.0	100	94	92	1	9.1	0.6
Enrofloxacin	50	10.5	33.8	33.7	114	97	93	0.992	103.8	6.3
Marbofloxacin	50	19.8	26.1	23.8	114	103	97	0.999	94.2	6.3
Difloxacin	2	13.0	21.8	18.2	99	96	95	0.996	3.4	0.3

Danofloxacin	30	21.7	27.3	20.2	117	109	101	0.996	59.3	3.8
Sarafloxacin	5	23.6	29.1	25.0	98	97	93	0.992	9.6	1.3
Sulfonamides										
Dapsone	5	17.0*	19.7	11.7	95	113	122	0.988	8.7	1.3
Sulfadiazine	50	11.5	6.6	5.2	106	91	92	0.993	59.9	6.3
Sulfamethoxazole	50	15.7	7.9	4.1	85	119	123	0.985	65.4	6.3
Sulfamethazine	50	12.1	11.7	13.4	102	102	94	1	69.6	6.3
Sulfadimethoxine	50	11.1	9.6	6.2	97	102	104	0.996	66.1	6.3
Sulfadoxine	50	11.2	7.9	11.0	103	101	105	0.998	63.0	6.3
Sulfaquinoxaline	50	8.2	5.7	9.3	99	108	104	0.999	60.0	6.3
Sulfachloropyridazine	50	21.9	7.1	6.5	86	106	113	0.996	62.3	6.3
Sulfamerazine	50	17.4	9.0	5.2	88	113	114	0.992	66.7	6.3
Sulfamethizole	50	6.3	6.1	8.6	126	87	81	0.975	58.7	6.3
Sulfamethoxypyridazine	50	8.8	6.4	5.5	103	112	107	0.999	61.7	6.3
Sulfamonomethoxine	50	10.9	6.9	4.0	93	110	107	0.997	62.5	6.3
Sulfamoxole	50	5.7	10.3	10.4	123	90	87	0.977	65.2	6.3
Sulfapyridine	50	22.6	11.9	9.6	87	112	114	0.994	71.9	6.3
Sulfisoxazole	50	38.0*	36.0*	36.5	90	107	106	0.994	117.6	6.3
Sulfathiazole	50	13.7	11.0	7.8	94	107	107	0.999	69.3	6.3
Pyrimidine										
Trimethoprim	25	17.3	14.4	10.1	107	111	112	0.998	38.1	3.1
Tetracyclines										
Tetracycline	100	15.9	11.1	12.4	104	82	84	0.989	129.9	12.5
Doxycycline	100	20.8	10.2	10.7	92	90	92	0.996	130.1	12.5
Oxytetracycline	100	21.1	12.1	9.6	103	92	95	0.993	136.4	12.5
Chlorotetracycline	100	26.7	15.9	13.0	129	80	73	0.976	141.8	12.5

(Continued)

TABLE 12.3. *(Continued)*

Compound	VL (µg/L)	%RSD Within-Lab Reproducibility			Accuracy (%)			(r²)	CCβ	LOQ
		0.5VL	1VL	1.5VL	0.5VL	1VL	1.5VL			
Nitroimidazolen										
Ipronidazol	5	35.6	39.6	33.5	165	186	181	0.996	17.1	0.6
Hydroxy-ipronidazol	5	23.9	14.5	13.5	119	114	112	1	7.7	1.3
Tranquillizers										
Azaperol	1	20.1	10.2	7.0	89	116	115	0.949	1.4	0.3
Azaperon	1	16.6	12.3	9.2	86	102	101	1	1.4	0.1
Propionylpromazine	1	39.0*	58.0*	50.0*	94	98	106	0.985		0.1
Acetopromazine	1	26.0*	26.0*	31.0*	79	97	98	0.993		0.1
Xylazine	1	15.4	12.0	8.6	79	101	105	0.996	1.4	0.1
Haloperidol	1	6.4	6.3	9.7	89	96	96	0.999	1.2	0.1
Chloropromazine	2	67.0*		85.0*	133		189	0.803		0.3
Carazolol	1	15.0	7.0	9.8	86	112	114	0.996	1.3	0.1
NSAIDs										
Piroxycam	50	9.5	9.5	7.1	98	112	110	0.999	67.4	6.3
Propyphenazone	20	7.3	7.0	9.4	123	105	90	0.997	24.8	6.3
Indoprofen	20	15.6	10.8	8.7	91	107	110	0.996	27.6	6.3
Tolmetin	50	16.8	13.7	14.3	103	115	115	0.997	75.9	6.3
Ketoprofen	50	39.0	13.9	12.7	101	120	114	0.995	77.3	6.3
Naproxen	50	38.0*	13.8	14.3	110	114	106	1	75.8	12.5
Fenbufen	50	20.4	9.4	8.9	103	106	104	1	66.3	6.3
Carprofen	50	15.0*	14.6	16.8	100	104	107	0.999	75.0	6.3
Diclofenac	50	24.1	9.2	13.4	80	105	108	0.999	65.8	6.3

	VL							r^2		
Niflumic acid	50	4.0*	9.0	9.7	98	102	103	—	65.0	6.3
Phenylbutazone	25	51.0*	55.0*	52.0*	97	135	166	0.954	95.4	6.3
Flufenamic acid	50	59.6*	14.5	11.9	47	87	93	0.986	70.8	6.3
Mefenamic acid	50	32.0*	15.4	15.0	58	97	99	0.962	74.6	6.3
Meclofenamic acid	50	21.0*	23.9	21.2	67	81	87	0.996	81.8	6.3
Isopyrin	50	36.0*	34.0*	28.0*	126	119	108	0.857	684.7	25.0
Isoxicam	50	9.6*	5.0	9.3		114	111	—	59.4	6.3
Tenoxicam	50	17.8	11.7	8.8	88	107	106	0.996	70.5	6.3
Sulindac	50	17.1	12.6	10.8	100	111	111	0.997	73.0	6.3
Indomethacin	50	29.0*	35.5	32.9	99	86	83	0.994	100.0	6.3
Flunixin	50	6.3	6.5	12.6	115	99	100	0.997	60.6	6.3
Ionophores										
Monensin Na	1	28.9	20.7	32.8	126	119	117	0.949	1.8	0.5
Salinomycine Na	1	37.0*	35.0*	37.0*	112	120	113	0.975	3.7	1.0
Narasin Na	1	40.0*	28.4	38.0*	97	93	126	0.988	1.9	0.5
Maduramycine Na	1	27.7	15.0	19.6	98	100	101	0.999	1.5	0.3
Amphenicols										
Florphenicol NH$_4$	5	32.0*	28.5	18.6	131	122	108	—	10.7	1.3
Thiamphenicol NH$_4$	50	16.0	15.1	10.6	99	115	112	0.993	78.5	6.3

VL = validation level (see also text)
r^2 = calibration curve correlation coefficient
*Validation study based on n = 14 (2 validation days); all other results are based on n = 21 (3 validation days).
Adapted from [23].

A

Repeatability

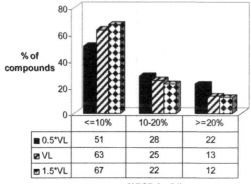

	<=10%	10-20%	>=20%
■ 0.5*VL	51	28	22
▨ VL	63	25	13
▨ 1.5*VL	67	22	12

%RSD (n=21)

B

Reproducibility

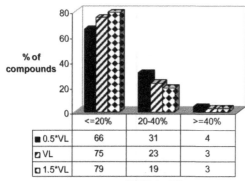

	<=20%	20-40%	>=40%
■ 0.5*VL	66	31	4
▨ VL	75	23	3
▨ 1.5*VL	79	19	3

%RSD (n=21)

C

Accuracy

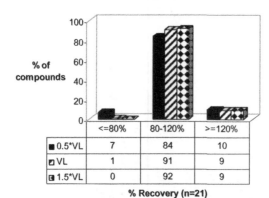

	<=80%	80-120%	>=120%
■ 0.5*VL	7	84	10
▨ VL	1	91	9
▨ 1.5*VL	0	92	9

% Recovery (n=21)

Figure 12.2.   Summary of (a) repeatability, (b) reproducibility, and (c) accuracy results for 101 veterinary drugs.

slightly better results are obtained for the later. Overall the repeatability results are acceptable for a quantitative multi-compound screening method.

From the within-lab reproducibility results presented in Figure 12.2b and Table 12.3 it is concluded that more than 75% of all compounds shows reproducibility results RSD < 20% at the 1 * VL. There are four compounds with unacceptable high (>40% at 1 * VL and/or 1.5 * VL) reproducibility results. These results are obtained for three tranquillizers propionylpromazine, acetompromazine and chloropromazine, and for the non-steroidal anti-inflammatory drug (NSAID): phenylbutazone. However, it can be concluded that for more than 96% of the compounds the reproducibility results are satisfactory.

From the accuracy results presented in Figure 12.2c and Table 12.3 it is concluded that more than 80% of the compounds show accuracy of >80%. There are only a few compounds with accuracy results <80% or >120% even at the 1 * VL. Compounds with an accuracy >120% are levamisole, (hydroxy-) thiabendazole, lincomycin, ampicillin, ipronidazol, phenylbutazone, and florphenicol. The compound with an accuracy (at 1 * VL) <70% is not surprisingly the unstable nafcillin. For these compounds accurate quantification is only possible by applying a standard addition procedure.

Linearity, CCβ, and LOQ   The linearity was determined for a concentration range of 0–0.25–0.5–1–2 and 4 times the VL. On each validation day the calibration curves were constructed. From these data the regression coefficients (r^2) were calculated. The criterium for good linearity was $r^2 > 0.99$. Values of r^2 for more than 80% of all matrix matched calibration curves—concentrations ranged from 0.25 * VL to 4 * VL—are >0.99. Moreover, all curves show an r^2 of >0.9 with only one exception, lincomycine with an r^2 of 0.87.

The detection capability (CCβ) at the VL was determined as follows: The CCβ = CCα + 1.64 * s.d.$_{(at VL)}$ in which CCα = VL + 1.64 * s.d.$_{(at VL)}$.

After evaluation of the results presented in Table 12.3 it was concluded that for the compounds with 1 * VL between 1 and 50 μg/L the CCβ are all within the range of 1.2–100 μg/L. Four drugs, lincomycin, enrofloxacin, sulphisoxazole, and isopyrin, show CCβ values of >100 μg/L at 1 * VL of 50 μg/L. For the tetracyclines the 1 * VL are 100 μg/L and the CCβ range from 129.9 to 141.8 μg/L.

In addition to the CCβ the limit of quantification (LOQ) was determined by analyzing seven samples of milk fortified at the concentration level of 1/8 * VL. In case the signal/noise at this concentration was ≥6, 1/8 * VL was set as the LOQ, otherwise the next calibration level with signal/noise ≥6 was set as LOQ.

The LOQ levels for all compounds are lower than the defined 1 * VL. For some drugs the 1 * VL was set a little higher than the advised MRPL. Taking this into account only one compound, hydroxy-ipronidazol, could not be detected because the advised MRPL of 1 μg/L is slightly lower than the LOQ of 1.25 μg/L. When the LOQs are compared with the results presented by Kaufmann et al. [10] for urine samples (>90% of the compounds LOQ < 10 μg/L) it is concluded that in this respect the presented method for milk is even more sensitive because all compounds tested show LOQs of <7 μg/L. It has to be mentioned that for the presented method a SPE step was included whereas Kaufmann only used a dilution step.

12.3.2   Sample Analysis

A set of approximately a hundred samples of raw milk (collected at farmhouses during the spring of 2007) were analyzed in four series of 25 samples each. Each series of samples started and ended with the analysis of the matrix matched calibration standards. The samples were analyzed by the method and the experimental conditions as described in [23].

The criteria for acceptance of the analytical results were:

- Sensitivity check (signal noise at LOQ level ≥6)
- deviation of the "r" value for the two calibration curves ≤ 20%
- relative retention time of suspected analyte and reference standard within the tolerance interval of ±2.5%

In case of a suspected result (positive response) the sample was reanalyzed for the specific compound(s). A suspected result is obtained when the response measured for an MRL substance is at or above CCβ level and for all other substances at or above LOQ level. The confirmatory analysis is based on a LC–QqQ-MS method monitoring two product ions of the suspected compound(s) thereby fulfilling the criteria described by the guidelines [5] for confirmatory analysis.

The 100 milk samples were analyzed on four different days. For all days the sensitivity and linearity checks were satisfactory. None of the samples was marked as suspected for containing a veterinary compound. However, the included reference sample was tested positive for sulfamethazine (Figure 12.3). The technician did not know that this reference sample was included (blind sample). The reference sample was previously analyzed by an LC-QqQ-MS method for containing sulfonamides and was tested positive for sulfamethazine (86 μg/L assigned value 87.8 μg/L).

All the samples (except the reference sample) were previously screened negative for the antibiotics (tetracylines, aminoglycosides, macrolides, and penicillins) by using microbiological screening tests. Microbiological screening tests are group specific. The UPLC–TOF-MS method is compound specific (individual drugs are identified). Furthermore it is very easy to add a new drug even from another class of drugs to the UPLC–TOF-MS method and retrospective analysis for a specific compound is possible by simply reprocessing of the acquired data and construction of the extracted ion chromatogram.

Figure 12.4, upper chromatogram, shows a typical example of a full scan TIC chromatogram of a sample of milk. Figure 12.4, lower chromatogram, shows a full scan chromatogram of a sample of milk fortified with the set of 101 veterinary drugs plus two internal standards at the 1 * VL concentration. Hardly any differences can be observed between the blank and the fortified sample. However, looking at an example of a typical extracted ion chromatogram of the blank presented in Figure 12.5a and the fortified sample of milk (at 1 * VL) presented in Figure 12.5b the differences are clear. The extracted ions are a representative subselection of the compounds tested, including albendazole m/z 266.0963, thiabendazole m/z 202.0439, flumequine m/z 262.0879, norfloxacin m/z 320.1410, enrofloxacin m/z 360.1723, sulphadiazine m/z 251.0602, sulphamethoxazole m/z 254.0599, tetracycline m/z 445.1611, and diclofenac m/z 296.0250.

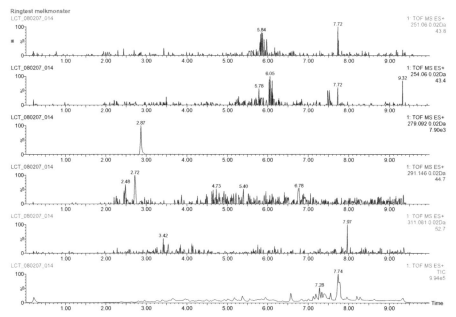

Figure 12.3.   UPLC–TOF-MS extracted ion chromatogram of check sample of milk with a selection of extracted ion chromatograms; The extracted ions, extraction windows 0.020 Da, are from top to bottom: sulfadiazine m/z 251.0603, Rt = 2.07 min; sulfamethoxazole m/z 254.0599, Rt = 3.44 min; sulfametazine m/z 279.0916, Rt = 2.87 min; trimethoprim m/z 291.145, Rt = 2.71 min; sulfadimethoxine m/z 311.0814, Rt = 4.26 min; sulfadimethoxine m/z 311.0814, Rt = 4.26 min. Track at the bottom of the figure is total ion current chromatogram.

12.4   DISCUSSION

In this section some of the important parameters for the successful use of UPLC–TOF-MS in multi-compound analysis of veterinary drug analysis are discussed.

12.4.1   Accurate Mass Measurements

Table 12.2 presents the accurate mass measurement data obtained for the multi-residue analysis of veterinary drugs in milk extracts [23]. In general for a TOF having a mass resolution of ~10,000 FWHM and an external calibration, a deviation of the measured accurate mass versus the calculated mass of 10 ppm is acceptable [24], especially when the low concentration levels (1–150 µg/L) are taken into account. From Table 12.2 it is concluded that the accurate measurements in a real sample extract are acceptable for more than 80% of the compounds studied. It is noted that 15 from the 19 compounds with a measured mass deviation >10 ppm elute with a retention time >5.5 min. Looking at the total ion current (TIC) chromatogram of a milk extract (Figure 12.4) it is obvious that beyond 5.5 min most of the matrix

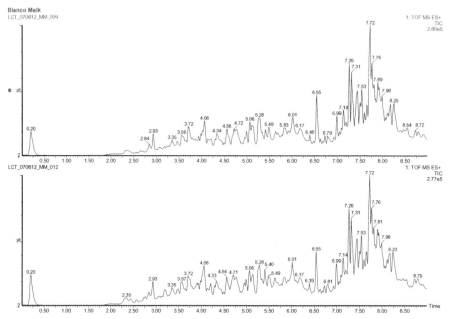

Figure 12.4 UPLC–TOF-MS full scan chromatogram of a blank milk sample (upper chromatogram) and a blank milk sample fortified with 101 veterinary drugs and 2 I.S (lower chromatogram) at the concentration of Validation Level (1*VL). For concentrations see Table 12.2; for analytical conditions see text.

compounds are eluting. For the analysis, 2 mL of milk is concentrated to a final volume of 500 μL from which 40 μL is injected corresponding with approximately 160 mg of matrix equivalent per injection. For the UPLC–TOF-MS method described in [17] only 0.4 μg of matrix was injected. Probably the unexpected low mass accuracy for compounds eluting above 5.5 min is due to elution of a large amount of matrix compounds [25]. However, the concentration step is necessary to detect the unauthorized compounds at the 1 μg/L level.

12.4.2 Extraction Window Width

From literature it is known that the selected width of the extraction window is important. When a very small extraction window width is selected there is possible change for false negative results [17]. Co-elution of isobaric compounds can result in significant deviations in exact mass measurements. The accurate mass assigned to a peak is the average of the accurate masses of the co-eluting isobaric peaks. Therefore it is possible that the average accurate mass is outside the extraction window width which is set for the specific compound of interest. The final result is a false negative observation [17]. When the extraction window is set at 0.05 Da (Figure 12.5a,b) all peaks are detected. When the extraction window is set at 0.01 Da as is shown in Figure 12.5c, still all the peaks are detected showing exactly the same

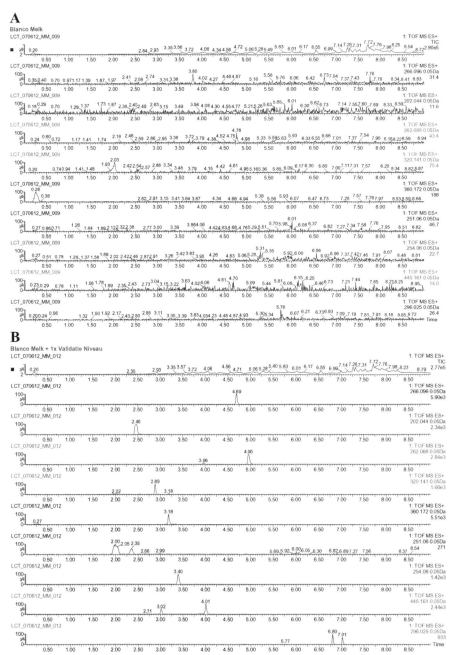

Figure 12.5. Typical example of a UPLC–TOF-MS extracted ion chromatogram of (a) blank sample of milk and (b) blank sample fortified at the 1*VL; The extracted ions, extraction window 0.050 Da, are from top to bottom:total ion current chromatogram; albendazole m/z 266.0963; thiabendazole m/z 202.0439; flumequine m/z 262.0879; norfloxacin m/z 320.1410; enrofloxacin m/z 360.1723; sulfadiazine m/z 251.0602; sulphamethoxazole m/z 254.0599; tetracycline m/z 445.1611; diclofenac m/z 296.0250; (c) is (b) but with an extraction window width of 0.010 Da (see text for more details). Adapted from [23].

C

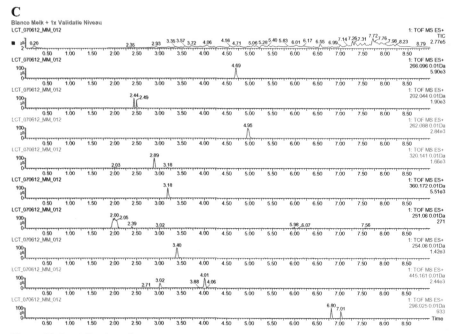

Figure 12.5. *(Continued)*

abundances, with thiabendazole (*m/z* 202.4399) as the only exception. The thiabendazole peak splits in two separate peaks at this low extraction window width. This
observation is possibly not due to an isobaric interfering compound co-eluting with
thiabendazole because in the blank sample no peak is detected. It is expected based
on the shape of the peak in Figure 12.5c, that this peak shows high intensity and
that detector is overloaded. When this occurs the DRE (dynamic range enhancement)
which is a technical device within the TOF, will electronically reduce the signal.
This functionality does not have influence on the acquired data. However, sometimes
this "correction" of signal results in a "dip" at the top of the peak. In summary the
peak is split probably due to the effect of the DRE.

The influence of the extraction window width is more obvious demonstrated
in Figure 12.6. Figure 12.6a presents the full scan TIC chromatogram of a sample
of milk. To find out if this sample contains oxolinic acid the extracted ion chromatogram of the ion *m/z* 262.0715 is constructed. Figure 12.6b presents the extracted
ion chromatogram with an extraction window of 1 Da. Two peaks are detected.
Reducing the width of the extraction window from 1 Da to 0.050 Da still two peaks
are detected one at the retention time of Rt = 3.96 min and one at Rt = 4.95 min. Only
when the extraction window is set at 0.010 Da one peak, oxolinic acid, at Rt = 3.96 min
is detected. The peak detected at Rt = 4.95 is flumequine with an accurate mass of
m/z 262.0879. In this typical example—also described by Kaufman for urine samples
[17]—the drugs can easily be separated based on the retention times, however when
this is not the case, for example when less selective LC is used, the two compounds

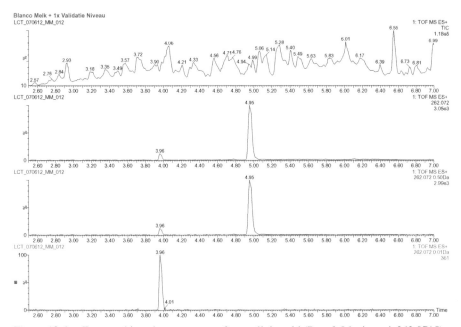

Figure 12.6. Extracted ion chromatograms for oxolinic acid (Rt = 3.96 min; m/z 262.0715) demonstrating the influence of the extraction window width; from top to bottom: TIC, extraction window of 1 Da, extraction window of 0.050 Da and lower track extraction window of 0.010 Da; at Rt = 4.96 min flumequine with m/z 262. 0879 elutes. Adapted from [23].

can only be separated from each other when the mass resolution of the TOF is high enough. The compounds differ by $262.0879 - 262.0715 = 0.0164$ Da. The mass resolution of the detector in combination with mass accuracy are important parameters for a reliable screening result. Some authors confuse sometimes the terms *extraction window width* and *mass resolution* [10]. These are two completely different things. The mass resolution is the power of the TOF-MS to separate two ions with only slightly different accurate masses. The extraction window width has nothing to do with MS resolution; it is just a parameter used to present the detected ions. In other words, it defines the borders of the masses of the ions which have to be extracted from the acquired data to construct the chromatogram.

12.4.3 LC Resolution

The importance of LC resolution in combination with high MS resolution and mass accuracy is demonstrated in Figure 12.7. This example shows the separation of albendazole sulfone (m/z 298.0862; Rt = 3.76 min) and hydroxy-mebendazole (m/z 298.1192; Rt = 3.58 min). At the specific 1 * VL concentration (albendazol sulfone 50 μg/L and hydroxy-mebendazole 2 μg/L) these compounds show significantly different detector responses. By using UPLC-TOF-MS these compounds can be

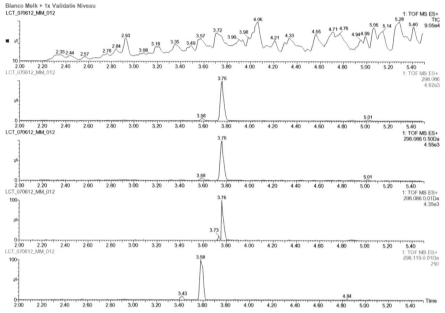

Figure 12.7. Extracted ion chromatograms for albendazole sulfone (*m/z* 298.0862; Rt = 3.76 min; VL = 50 µg/L); from top to bottom: TIC, extraction window of 1 Da, extraction window of 0.050 Da and extraction window of 0.010 Da; lower track is the extracted ion chromatogram for hydroxy-mebendazole (*m/z* 298.1192; Rt = 3.58 min; VL = 2 µg/L) extraction window is 0.010 Da.

separated and detected. When the extracted ion chromatogram is constructed for *m/z* 298.062 (width of 0.10 Da) only albendazole sulfone is monitored and when the extracted ion is *m/z* 298.1192 (width of 0.010 Da) only hydroxy-mebendazol is monitored. Co-elution of these two peaks, for example when regular LC had been used will definitely result in a false negative observation for hydroxy-mebendazol. The huge peak of albendazole sulfone will swallow the small peak of hydroxy-mebendazol. The influence of hydroxy-mebendazol on the average accurate mass of the "merged peak" will be limited so with an extraction window width of 0.050 Da, just one peak, albendazole sulfone, will be shown in an extracted ion chromatogram [25, 26].

From the examples given above it is concluded that UPLC–TOF-MS is a very powerful technique for multi-compound detection and in theory for an unlimited number of compounds [27].

12.5 EU CRITERIA

According to Nielen et al. [19] accurate-mass LC–TOF-MS screening of target residues and accurate-mass confirmation of known and identification of unknown,

TABLE 12.4.   Proposal for Additional LC–MS Criteria to Be Implemented in 2002/657/EC[5]

Function	Mass resolution (FWHM)	Mass accuracy (mDa)	Remarks
Screening	≥10,000	±50(window)	Relative retention time ≤2.5%
Confirmation	≥10,000	≤5	1.5 IPs per ion or product-ion; Minimum of 1 ion ratio; Relative retention time ≤2.5%
HR confirmation	≥20,000	≤5	2 IPs per ion or product-ion; Minimum of 1 ion ratio; Relative retention time ≤2.5%
MS/MS identification of unknowns	≥10,000	≤5	Confirm postulated structure by NMR and/or confirm accurate masses at mass resolution ≥70,000 (FWHM)

Adapted from [19].

residues will expand in the near future. Criteria for the confirmation of the identity are not described for this technique by Commission Decision 2002/657/EC. In this EU document high resolution mass spectrometry (HRMS) is defined as MS at a mass resolution of 10,000 according to the 10% valley definition. This definition originates from the double-focusing GC–HRMS. For modern instruments such as TOF-, FTICR-, and FT Orbitrap-MS the resolution is usually specified as full width at half maximum (FWHM). A resolution of 10,000 according to the 10% valley definition corresponds to approximately. 20,000 FWHM. According to [5], the criteria for HRMS are as follows: 2.0 IP are earned for each ion or precursor ion, and 2.5 IPs for each product ion. No criteria for mass accuracy are described. Nielen et al. demonstrated that for confirmatory analysis using TOF-MS the mass spectrometric resolution has significant influence on the mass accuracies obtained. Insufficient mass resolution yields incorrect accurate mass data and might cause false compliant screening and/or confirmatory results. Based on this experience the authors proposed additional criteria for Commission Decision 2002/657/EC. Table 12.4 presents these proposed criteria. They suggest to collect three TOF-MS ions in order to earn 4.0 IPs. Only when using high-resolution techniques the use of only one parent plus one product ion is acceptable for confirmatory purposes. Furthermore it is proposed that a single TOF-MS analysis can earn the same number of IPs as a QqQ-MS analysis operated in the MRM mode which is a very realistic approach.

On the basis of the experimental work it was demonstrated that for the confirmation of postulated structures of unknown hormones and β-agonists very high mass resolution is required. The proposed resolution of ≥70,000 (FWHM) assures that reliable elemental compositions of product ions differing in one CO, C_2H_4, or N_2 substructure can be obtained up till m/z 400.

12.6 CONCLUSIONS

There is an increasing interest in methods for the simultaneous analysis of various classes of veterinary drugs. Such multi-compound analyses which, in one run, can deal with more than 100 compounds are only possible by using full mass scan MS techniques like TOF-MS. Several examples have already been published, with UPLC–TOF-MS probably being the most powerful measurement tool as is also demonstrated by the present study. The use of UPLC is especially powerful when biological extracts have to be screened because the additional LC selectivity compensates for the lack of selectivity in comparison with a MS/MS option of a QqQ-MS.

In the past, a LC–QqQ method (utilizing selected MRM transitions) was the starting point for method development. Today, in theory all compounds can be measured by full scan MS. The starting point now is no longer the detection conditions used, but the sample material. Starting with for example milk all relevant veterinary drugs—irrespective of the class they belong to—can be detected in one (UP)LC–TOF-MS analysis, and the most important part of the method development is the generic extraction of the compounds of interest out of the matrix. This will no doubt cause an increased interest in a variety of modern sample pretreatment techniques. The sample pretreatment step becomes even more of interest when also other organic contaminants like residues of pesticides, mycotoxins and or plant toxins are included in the comprehensive screening method.

In summary, the UPLC–TOF-MS combination offers unsurpassed performance for screening purposes, it can also effectively provide concentration values, and is accurate enough to differentiate between positive and negative samples or drug concentrations below or above the MRL/MRPL.

Finally it has to be mentioned that the confirmatory analysis of the suspected samples have to be done by the use of an MS/MS technique because criteria for confirmation of the identity of drug by TOF-MS are not included yet in the guidelines of the EU.

The introduction of new MS techniques like TOF also has consequences for the EU criteria defined for confirmation of compound identity. Both mass resolution and mass accuracy data are of influence for such identity confirmation and both parameters should preferably be included in the revised version of Commission Decision 2002/657/EC.

REFERENCES

1. Official Journal of the European Union, L224 18 August 1990, Council Regulation 2377/90/EC of 26 June 1990 laying down a community procedure for the establishment of maximum residue limits of veterinary medicinal products in foodstuffs of animal origin, Brussels, Belgium, **1990**.
2. Official Journal of the European Union, L125 23/05/1996, Council Directive 96/22/EC of 29 April 1996 concerning the prohibition on the use in stockfarming of certain substances having a hormonal or thyrostatic action and of beta-agonists, and repealing Directives 81/602/EEC, 88/146/EEC and 88/299/EEC, pp 3–9, Brussels, Belgium, **1996**.

3. Official Journal of the European Union, L125, 23/05/1996, Council Directive 96/23/EC of 29 April 1996 on measures to monitor certain substances and residues thereof in live animals and animal products and repealing Directives 85/358/EEC and 86/469/EEC and Decision 89/187/EEC and 91/664/EEC, pp 10–32, Brussels, Belgium, **1996**.

4. Official Journal of the European Communities L224, of 18 August 1990, Council Regulation 2377/90/EC; consolidated version of the Annexes I to IV updated up to 22.12.2004 obtained from www.emea.eu.int.

5. Official Journal of the European Communities L221, 8–36. Commission Decision (2002/657/EC) of 12 August 2002, Brussels, Belgium, 2002.

6. Bibliographic databanks, Toxline and Current Contents, 2005–2006–2007.

7. Stolker, A.A.M.; Brinkman, U.A.Th., Analytical strategies for residue analysis of veterinary drugs and growth-promoting agents in food-producing animals—a review, *J. Chromatogr. A*, **2005**, 1067, 15–53.

8. Stolker, A.A.M.; Zuidema, T.; Nielen, M.W.F., Residue analysis of veterinary drugs and growth-promoting agents, *Trends Anal. Chem.*, **2007**, 26, 967–979.

9. Bailac, S.; Barron, D.; Barbosa, J., New extraction procedure to improve the determination of quinolones in poultry muscle by liquid chromatography with ultraviolet and mass spectrometric detection, *Anal. Chim. Acta*, **2006**, 580, 163–169.

10. Vinci, F.; Fabbrocino, S.; Fiori, M.; Serpe, L.; Gallo, P., Determination of fourteen non-steroidal anti-inflammatory drugs in animal serum and plasma by liquid chromatography/mass spectrometry, *Rapid Commun. Mass Spectrom.*, **2006**, 20, 3412–3420.

11. Samanidou, V.F.; Nikolaidou, K.I.; Papadoyannis, I.N., Development and validation of an HPLC confirmatory method for the determination of tetracycline antibiotics residues in bovine muscle according to the European Union regulation 2002/657/EC, *J. Sep. Sci.*, **2005**, 28, 2247–2258.

12. Hebestreit, M.; Flenker, U.; Fussholler, G.; Geyer, H.; Güntner, U.; Mareck, U.; Piper, Th.; Thevis, M.; Ayotte, Ch.; Schänzer, W., Determination of the origin of urinary norandrosterone traces by gas chromatography combustion isotope ratio mass spectrometry, *Analyst*, **2006**, 131, 1021–1026.

13. Noppe, H.; Verheyden, K.; Gillis, W.; Courtheyn, D.; Vanthemsche, P.; De Brabander, H.F., Multi-analyte approach for the determination of ng L^{-1} levels of steroid hormones in unidentified aqueous samples, *Anal. Chim. Acta*, **2007**, 586, 22–29.

14. Verdon, E.; Couedor, P.; Sanders, P., Multi-residue monitoring for the simultaneous determination of five nitrofurans (furazolidone, furaltadone, nitrofurazone, nitrofurantoine, nifursol) in poultry muscle tissue though the detection of their five major metabolites (AOZ, AMOZ, SEM, AHD, DNSAH) by liquid chromatography coupled to electrospray tandem mass spectrometry—In-house validation in line with Commission Decision 657/2002/EC, *Anal. Chim. Acta*, **2007**, 568, 336–347.

15. Granelli, K.; Branzell, C., Rapid multi-residue screening of antibiotics in muscle and kidney by liquid chromatography-electrospray ionization-tandem mass spectrometry, *Anal. Chim. Acta*, **2007**, 586, 289–295.

16. Yamada, R.; Kozono, M.; Ohmori, T.; Morimatsu, F.; Kitayama, M., Simultaneous determination of residual veterinary drugs in bovine, procine, and chicken muscle using liquid chromatography coupled with electrospray ionization tandem mass spectrometry, *Biosci. Biotechnol. Biochem.*, **2006**, 70, 54–65.

17. Kaufmann, A.; Butcher, P.; Maden, K.; Widmer, W., Ultra-performance liquid chromatography coupled to time of flight mass spectrometry (UPLC-TOF): A novel tool for multiresidue screening of veterinary drugs in urine, *Anal. Chim. Acta*, **2007**, 586, 13–21.

18. Muñoz, P.; Blanca, J.; Ramos, M.; Bartolomé, M.; García, E.; Méndez, N.; Gomez, J.; De Pozuelo, M.M., A versatile liquid chromatography-tandem mass spectrometry system for the analysis of different groups of veterinary drugs, *Anal. Chim. Acta*, **2005**, 529, 137–144.

19. Nielen, M.W.F.; van Engelen, M.C.; Zuiderent, R.; Ramaker, R., Screening and confirmation criteria for hormone residue analysis using liquid chromatography accurate mass time-of-flight, Fourier transform ion cyclotron resonance and orbitrap mass spectrometry techniques, *Anal. Chim. Acta*, **2007**, 586, 122–129.

20. Hernando, M.D.; Mezcua, M.; Suárez-Barcena, J.M.; Fernández-Alba, A.R., Liquid chromatography with time-of-flight mass spectrometry for simultaneous determination of chemotherapeutant residues in salmon, *Anal. Chim. Acta*, **2006**, 562, 176–184.

21. Stubbings, G.; Tarbin, J.; Cooper, A.; Sharman, M.; Bigwood, T.; Robb, P., A multi-residue cation-exchange clean up procedure for basic drugs in produce of animal origin, *Anal. Chim. Acta*, **2005**, 547, 262–268.

22. Bogialli, S.; Di Corcia, A., Matrix solid-phase dispersion as a valuable tool for extracting contaminants from foodstuffs, *J. Biochem. Biophys. Meth.*, **2007**, 70, 163–179.

23. Stolker, A.A.M.; Rutgers, P.; Oosterink, E.; Lasaroms, J.J.P.; Peters, R.J.B.; van Rhijn, J.A.; Nielen, M.W.F., Comprehensive screening and quantification of veterinary drugs in milk using UPLC-TOF-MS, Bioanal, *Anal. Chem.*, **2008**, 391, 2309–2322.

24. Ojanperä, S.; Pelander, A.; Pelzing, M.; Krebs, I.; Vuori, E.; Ojanperä, I., Isotopic pattern and accurate mass determination in urine drug screening by liquid chromatography/time-of-flight mass spectrometry, *Rapid. Commun. Mass Spectrom.*, **2006**, 20, 1161–1167.

25. Calbiani, F.; Careri, M.; Elviri, L.; Mangia, A.; Zagnoni, I., Matrix effects on accurate mass measurements of low-molecular weight compounds using liquid chromatography-electrospray-quadrupole time-of-flight mass spectrometry, *J. Mass Spectrom.*, **2006**, 41, 289–294.

26. Laures, A.M.F.; Wolff, J.C.; Eckers, Ch.; Borman, Ph.J.; Chatfield, M.J., Investigation into the factors affecting accuracy of mass measurements on a time-of-flight mass spectrometer using Design of Experiment, *Rapid. Commun. Mass Spectrom.*, **2007**, 21, 529–535.

27. Ibáñez, M.; Sancho, J.V.; McMillan, D.; Rao, R.; Hernández, F., Rapid non-target screening of organic pollutants in water samples by ultra performance liquid chromatography-time of flight mass spectrometry, *TRAC Trends Anal. Chem.*, **2008**, 27, 481–489.

CHAPTER 13

ACCURATE MASS MEASUREMENTS WITH A REFLECTRON TIME-OF-FLIGHT MASS SPECTROMETER AND THE DIRECT ANALYSIS IN REAL TIME (DART) INTERFACE FOR THE IDENTIFICATION OF UNKNOWN COMPOUNDS BELOW MASSES OF 500 DA

O. David Sparkman, Patrick R. Jones, and Matthew Curtis

Pacific Mass Spectrometry Facility, Department of Chemistry, College of the Pacific, University of the Pacific, Stockton, California

ACCURATE MASS measurement as provided by today's time-of-flight (TOF) mass spectrometers allows for determination of elemental compositions of small molecules. However, elemental composition alone does not necessarily provide a structure. Ease of obtaining mass spectral data has been greatly facilitated by the development of the direct analysis in real time (DART) ionization source from JEOL USA, Inc. This ion source, in combination with the accurate mass assignments for ions from the modern reflectron TOF instrument, provides for very quick analysis of unknowns. Combining this technology with hydrogen/deuterium exchange in the DART ion source has proven very valuable in allowing for the determination of not only the elemental composition of an analyte but also its structure. This chapter illustrates how the DART ion source, the reflectron TOF

Liquid Chromatography Time-of-Flight Mass Spectrometry: Principles, Tools, and Applications for Accurate Mass Analysis, Edited by Imma Ferrer and E. Michael Thurman
Copyright © 2009 John Wiley & Sons, Inc.

mass spectrometer, and hydrogen/deuterium exchange are used to identify unknown substances at the same time as providing an introduction to how the DART ion source functions.

13.1 INTRODUCTION

The development of the reflectron time-of-flight (TOF) mass spectrometer revolutionized the field of mass spectrometry. Ever since these instruments began to appear as stand alone mass spectrometers with electrospray ionization interfaces in 1997 [1], their continued commercial improvements have brought the area of accurate mass analyses to the average analyst. No longer is accurate mass the privy of double-focusing instruments with all of their idiosyncrasies and need for highly skilled operators. The atmospheric pressure ionization (API) TOF mass spectrometer, as commercialized by Waters, JEOL, Agilent Technologies, Bruker Daltonics, and other manufacturers, has proved to be easy to use, reliable, and indispensable in many laboratories. The TOF mass spectrometer has also proved to be a major contributor to a number of successful tandem-in-space instruments such as the quadrupole-TOF instruments from Waters, Applied BioSystems, Agilent Technologies, and Bruker Daltonics. Some modern TOF mass spectrometers are advertised with resolving powers (R) as high as 20,000.

One major advantage the TOF instrument has over the double-focusing mass spectrometer for mass accuracy measurements is in the area of calibration. In order to achieve accurate mass assignments with the double-focusing instrument, it is necessary to have the peaks used for mass calibration in the same spectrum as the peaks whose accurate masses are being assigned. With the TOF instrument, good assignments can be achieved when the mass calibration ions are in the spectra in the same data file. Spectra that contain peaks representing the calibration ions can be acquired at the beginning or the end of a data file.

Perhaps the most remarkable feature of the API TOF instrument is its propensity to provide data on a day-to-day basis with minimal maintenance and little or no tuning. To obtain the same performance, the double-focusing mass spectrometer requires daily tuning at a minimum and, in some cases, tuning before each acquisition. The major maintenance issue with the TOF mass spectrometer is regular changing of the oil in the mechanical pump. In all fairness, it should be pointed out that most experiences with double-focusing instruments involve ionization inside the mass spectrometer, such as electron ionization (EI) and fast atom bombardment (FAB), both of which contributed to contamination, necessitating frequent tuning. The TOF mass spectrometer, on the other hand, is primarily used with an atmospheric pressure interface, which means the ions are not formed in the vacuum region of the instrument. Today, the use of the API TOF mass spectrometer for accurate mass measurements can be considered as simple and straightforward as unit-resolution acquisition as the transmission quadrupole mass filter (QMF) or the quadrupole ion trap (QIT) mass spectrometer. In addition, just like both of these instruments, the level of productivity of the TOF mass spectrometer can be considered to be very high. It is likely that in the not-too-distant future, the TOF mass spectrometer will become as ubiquitous as the QMF is today.

13.2 INFORMATION CONTENT

Accurate mass measurements can provide an elemental composition. This ability, however, is limited by the mass accuracy and the m/z value of the ion. As pointed out by Michael Gross [2], a single-charge ion with m/z 118 was free of competing elemental composition within ±34 ppm "when valence rules and candidate compositions of C_{0-100}, H_{3-74}, O_{0-4}, and N_{0-4} are considered." At less than ±5 ppm, following the same rules, a single-charge ion with an m/z value of 500 yielded five elemental compositions. A single-charge ion of m/z 750.4 requires a mass accuracy of ±0.0018 ppm for an unambiguous elemental composition. If the mass accuracy can yield an unambiguous elemental composition, the analyte identification can still be elusive. Take, for example, the elemental composition $C_{12}H_{18}Cl_2N_2O$. With a nominal mass of 276 Da, this composition could easily be obtained using data from a TOF mass spectrometer. The rings-plus-double-bonds count for this formula is 5. The number of possible configurations that could be deduced from such information is larger than can be easily conceived. In order to identify this analyte (which may only be capable of producing ions representing the intact species because it is not amenable to EI) will require more than just an unambiguous elemental composition of the intact molecule. If MS/MS is available and the ion with m/z 276 is isolated, fragment ions that may be obtained could supply the necessary additional information, or observed neutral losses (dark matter) may allow further elucidation of the analyte. Mass accuracy is often not available for these fragmentation ions unless something like a hybride instrument such as a tandem transmission quadrupole in line with a TOF is available; therefore, dark matter consisting of 43 Da could represent the loss of a propyl radical or $H_3C^•C{=}O$ radical. The data may not reveal what functionalities are associated with hetero atoms such as oxygen and nitrogen. Accurate mass assignments of fragments can be obtained with a reflectron TOF instrument but care must be taken because the precursor ion of interest must be isolated as a function of chromatography.

Derivatization can provide some information, but this is time consuming; and as in the case of the $C_{12}H_{18}Cl_2N_2O$ analyte, the results may be very complex because this analyte has three derivatizable sites.

This dilemma of limited information has plagued atmospheric pressure ionization mass spectrometry since its commercialization. The same dilemma threatened to be the bane of DART (Direct Analysis in Real Time) [3] when it was developed. It was soon discovered [4] that in situ hydrogen/deuterium exchange (H/D) could be used to determine the presence, number, and nature of functional groups on a target molecule. Using the accurate mass measurement that can be achieved with the TOF mass spectrometer and in situ H/D exchange and techniques of data mining described by Cody [5], an illustration of the identification of an unknown is presented.

13.3 DART INTERFACE FOR SAMPLING AND IONIZATION

The DART interface, invented by Robert B. Cody and James A. Laramée (JEOL USA) and patented in 2006, functions as both an ion source and a sampling device. Samples are held in the space between the exit from the DART source and the atmospheric pressure inlet to a mass spectrometer. The DART interface has been mated with a number of different mass spectrometers including TOF, triple quadrupole, and QIT instruments. Ionization/desorption is a result of a heated stream of gas (helium or nitrogen) rich in metastable species directed towards the atmospheric pressure inlet of the mass spectrometer. The temperature of the gas stream is operator controlled and can vary from room temperature to over 400 °C. The flow rate is set by the design of the device and the distance between the interface exit and the mass spectrometer inlet is variable by a few centimeters. Analyses are performed by holding a sample in the space between the DART source and the atmospheric pressure inlet of the mass spectrometer in the gas stream. This allows for the analysis of whatever is on the surface of the sample. In the case of liquids, a small amount is placed on the outside of a clean glass capillary melting point tube, on a piece of filter paper, or on ceramic paper. Other substances already on the surface of such a carrier are directly analyzed.

The exact nature of the ionization process is still speculative. In the sample gap (the space between the ion source and the entrance to the mass analyzer), the ionization of the analytes is due to a process initiated by the interactions between a stream of helium or nitrogen metastable atoms and water vapor in air without requiring the presence of solvents or significant electrical potentials. The DART produces helium metastable atoms as helium flows through the cartridge. Figure 13.1 shows a simplified illustration of the DART interface. Cody and Laramée provide a more detailed explanation of the ion source [3]. The helium flows into the back of the cartridge and then into the discharge chamber through the needle electrode. The potential between the needle and disc electrode is approximately 3000 volts. The glow discharge, or plasma, that is produced is made up of charged species and helium metastable atoms. The neutral helium metastable atoms arise from the excitation of neutral helium atoms such that the two electrons are in higher energy states. For helium, the state of interest for the DART source is the He triplet S state symbolized by He^3S (the symbol for excited-state helium where the spin multiplicity is 3 to give a triplet state) [6]. The triplet S state is considered metastable because the quantum physics involved requires a spin flip, which has a very low probability for the isolated atom. However, on encounter with an analyte such as an organic molecule, the energy stored in the metastable helium may be released [6], making approximately 19.8 eV of energy available for desorption and ionization.

In order to eliminate all charged species from the plasma allowing only the helium metastable atoms to remain, the plasma flows through a second disc electrode. Before the stream of helium metastable atoms flows into the sample gap, it flows through a molybdenum grid electrode; this final grid electrode, electrode 2, acts as an ion repeller preventing newly formed ions produced in the sample gap from flowing back into the source and prevents ion–ion recombination [1]. Elec-

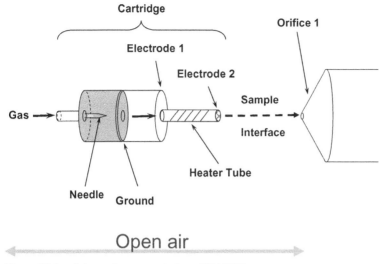

Figure 13.1.   Schematic representation of DART ion source.

trode 2 is covered with a ceramic insulating cap to protect samples and operators from high voltage. The position of the ceramic cap can be manipulated by a knob at the back of the encasement as well as two bolts, one located on top and one to the side of the encasement. They allow adjustments for optimal positioning of the helium stream relative to Orifice-1 of the *AccuTOF*, the entrance to the mass analyzer.

Based on observations of the different types of ions that are formed, there are many ionization processes taking place in the interface; however, the ions encountered most often in the positive mode are protonated analyte molecules. The helium metastable atoms form molecular ions of water through Penning ionization; a process by which the helium metastable atoms (He^3S) interact (collisions) with water molecules in the air and transfer energy to the water molecules to produce energetically activated unstable molecules and ground state helium atoms, He^1S. In order to stabilize, the water molecules eject thermal electrons, e^-, to produce water molecular ions, $H_2O^{+\bullet}$ (Equation 1.1).

$$He(^3S) + H_2O \rightarrow H_2O^{+\bullet} + He(^1S) + e^- \tag{1.1}$$

The water molecular ions then interact with other water molecules in the air to produce protonated water molecules, H_3O^+, and hydroxyl radicals, $OH^\bullet$ (Equation 1.2).

$$H_2O^{+\bullet} + H_2O \rightarrow H_3O^+ + OH^\bullet \tag{1.2}$$

The protonated water molecules continue to interact with other water molecules (n = number of water molecules) and soon produce protonated water clusters containing m numbers of water molecules where $m = n + 1$ (Equation 1.3).

$$H_3O^+ + nH_2O \rightarrow \left[(H_2O)_m + H^+\right] \tag{1.3}$$

The protonated water clusters interact with the analyte, A, in the open-air source. If the analyte has a higher proton affinity than water, the analyte will become protonated, AH^+, enabling detection by a mass spectrometer (Equation 1.4).

$$[(H_2O)_m + H]^+ + A \rightarrow AH^+ + mH_2O \tag{1.4}$$

The ionization mechanism outlined produces a relatively uncomplicated mass spectrum. The ionization mechanism does not produce multiple-charge ions as sometimes observed in electrospray (ES). When analyzing analytes in solution, solvent adducts are sometimes observed to be produced by a mechanism similar to clustering in atmospheric pressure chemical ionization (APCI). When acetonitrile is used as a solvent, $M(CH_3CN)H^+$ ions have been observed. At times, where a certain type of adduct is desired, dopants can be used. For instance, the introduction of ammonia into the sample gap will produce ammonium adduct ions, $[M + NH_4]^+$, along with protonated molecules. All ions are produced in the absence of significant potentials producing minimal sample degradation.

In addition to protonated molecules, molecular ions of some analytes have been observed to be formed through Penning ionization. Negative ions can be formed by the electron capture of the thermal electrons by oxygen molecules resulting from positive-ion formation of water molecules. The resulting negative-charge oxygen molecular ions then react with analyte molecules through an ion/molecule collision to produce negative-charge ions. Electrophilic analytes will form negative-charge molecular ions by the direct resonance electron capture of these thermal electrons.

One of the most important tunable parameters of the DART ion source is the temperature to which the He gas is heated. Whereas temperatures above a compound's boiling point may aid desorption of the compound, higher temperatures may degrade the compound or the matrix in which it is found. For instance, when an unknown captured on filter paper was analyzed, careful consideration went into choosing the analysis heater temperature because the filter paper tended to singe at temperatures higher than 100 °C and began to produce pyrolysis products. The need to protect the matrix showed that ionization can occur at lower temperatures; temperature of the He gas acting as an agent to volatilize the analyte is not the only determinant of ionization of a particular compound. The mechanisms of all the different aspects of DART ionization are still undermined. It may be that ionization is taking place of analytes in the condensed phase and these ions are then desorbed into the gas phase. Many compounds are detected at lower temperatures including when the He gas heater was turned off and allowed to cool until the temperature readout was 18 °C.

13.4 ILLUSTRATIVE EXAMPLE OF A DART ANALYSIS USING A TOF MASS SPECTROMETER

A complex aqueous solution was found to perform differently when formulated with water from different sources. The solution failed to perform properly when formulated from one water source whereas performance was fine when another water-source was used. A quantity of water was filtered, and the filter was analyzed to determine if any organic compounds had been trapped which could be responsible for the observed solution-failure.

The filters can be described as a filter paper assembly encased in plastic to form a cylinder 0.50 m in length with a 3.5-cm inner diameter and a 7.0-cm outer diameter. The entire filter, encasing and all, was sliced using a band saw in order to access the filter paper assembly. The plastic encasement was discarded, leaving a cylinder of the accordion-folded filter paper assembly. Samples for analysis were cut so that two folds were able to be analyzed; one fold from the "inside" of the original filter (Side 1) and one from the "outside" of the original filter (Side 2). The filter paper assembly is composed of three layers; the first and third layers are constructed of large fibers loosely woven, whereas the middle layer appeared to be a very pliable shiny paper. The middle layer was determined to be where the unknown would most likely collect based on analyses of all three layers. Forceps were used to hold the filter paper in the DART-*AccuTOF* sample gap during analysis. Each fold was held perpendicular to the He stream.

The used filter was wet with residual water resulting from its use. It was analyzed as received. Because this used filter was no longer encased in the plastic housing, the filter did dry on standing over a period of time. Analyses of the subsequently dried filters did not reveal any mass spectral differences from those obtained with the wet filters.

13.4.1   Experimental

The PALL Aquaflow® filters with pore size $0.45 + 0.2\,\mu m$ were analyzed at 100 °C. Initial experiments with samples from the unused filters at ambient temperature (DART heater off), 100 °C, and 200 °C with and without $NH_{3(aq)}$ in the positive-ion mode indicated that the optimal temperature was 100 °C. At 200 °C, the filter paper showed signs of charring. As previously pointed out, the possibility of pyrolysis products from the matrix should be avoided.

The *m/z* scale was calibrated using polyethylene glycol (PEG) dissolved in methanol introduced into the DART-*AccuTOF* sample gap on the outer surface of a glass melting point capillary; mass spectral peaks 44 *m/z* units apart are produced over a wide spectral range. PEG 200 is added to the calibration mixture in order to acquire better signal and thus better calibration at the lower end of the *m/z* scale. The best heater temperature to use for the PEG calibrant is approximately 200 °C, which gives ample signal across the mass spectral range; at lower temperatures, mass spectral peaks at higher *m/z* values lack sufficient intensity for calibration. At higher temperatures, mass spectral peaks at higher *m/z* values are observed and calibration is possible; however, good calibration at the lower end of the mass spectral range is sacrificed. In sample analyses where temperatures less than 200 °C are required, calibration can be done at the end of the acquisition following sample analyses. After all of the samples have been analyzed, the temperature of the heater is increased and the calibration data are acquired to the same data file that contains the data for the samples. Once the heater temperature readout has reached a desired level, a one-minute stabilization period is required before introducing the PEG into the sample

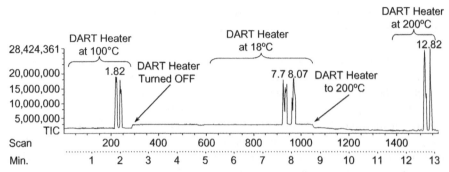

Figure 13.2. TIC of a data file showing the fluctuations that occur in the baseline as a result of the DART heater temperature changing.

gap to allow the baseline to stabilize. In an analysis that requires an extended acquisition time (i.e., over 30 minutes), calibrant should be introduced at the beginning and near the end of the acquisition to ensure good mass accuracy throughout the acquisition.

Changing the heater temperature of the DART is evident in the total ion current (TIC) display; the baseline changes due to fluctuations in temperature. In Figure 13.2, notice the rise of the baseline when the temperature is lowered, and the drop of the baseline when the temperature is increased.

Each sample is analyzed three times (replicates). If sample is a large sheet, three different spots on the sheet are analyzed to count as three replicates. Using data analysis software, the mass spectral data acquired from the samples and the blank are compared to determine the mass spectral peaks that are not associated with the blank or laboratory background.

After some experimentation, the DART-*AccuTOF* conditions shown in Table 13.1 were selected.

In order to produce in-source collisionally activated dissociation (CAD) fragmentation, Orifice-1 was also operated at 40 V and 60 V. The operating resolving power (R) of the *AccuTOF* is approximately 7000 (FWHM). Spectra of polyethylene glycol (PEG) were acquired for each data file to provide an internal accurate mass calibration.

The used filters were drenched in water as a result of their use; therefore, they were dried in an 80 °C oven for approximately 20 minutes prior to H/D exchange analysis. The drying process was carried out so that any H/D exchanges that occurred were only due to the unknown captured on the filter paper. To facilitate the H/D exchange, the filter papers were drenched with D_2O instead of using the nebulizer.

Once the unknowns were tentatively identified, authentic samples of dipropylene glycol (DPG) and diethylene glycol monoethyl ether (DEGE) were purchased from Sigma-Aldrich (St. Louis, MO). The disposable glass melting point capillaries were purchased from VWR (Brisbane, CA). The standard compounds DEGE and DPG were analyzed in their pure form, a liquid. In order to analyze the samples, the sealed end of the melting point capillaries were carefully dipped into the viscous

TABLE 13.1. Experimental Parameters

Helium flow	$4.2\,L\,min^{-1}$
He temperature	$100\,°C$
DART discharge needle voltage	5629 V in positive-ion mode
Orifice-1	*Operated at 20 V*
Acquisition rate	1 averaged spectrum saved every 0.470 sec with a delay time of 3 milliseconds and a data sampling interval of 0.5 ns per point without the use of data compression
Spectral acquisition range	*m/z* 50–1000

liquids and then placed in the helium gas stream of the DART ion source. The amount of sample analyzed was the amount that adhered to the surface of the glass melting point capillary (a volume estimated to be 1 μL).

After assigning accurate *m/z* values to the mass spectra peaks of interest, the *Elemental Compositions* (*ElComp*) software in the *MS Tools* suite (ChemSW, Inc., Fairfield, CA) is used to assign elemental compositions that are unique to the samples; these peaks will be considered to represent unknown compounds. Using the calculated elemental compositions and basic mass spectrometry interpretation skills, a determination of the number of "unknowns" present in the samples is made. For instance, if a pair of mass spectral peaks differ by 18 *m/z* units, these peaks may represent a protonated molecule, $[M + H]^+$, and an ion that results when a molecule of water is lost from the protonated molecule, $[M + H - H_2O]^+$.

The elemental compositions generated from the *ElComp* software are adjusted to those of molecules rather than protonated molecules and searched using the *Formula Search* feature of the *NIST Mass Spectral Search Program* (*MS Search*) to search against the *NIST05 Mass Spectral Database* (*NIST05*). When data are acquired in the positive-ion mode, one is subtracted from the elemental composition and the nominal mass of a H atom is added to the elemental composition if the data are acquired in the negative-ion mode.

It should be pointed out that there are other sources of listings of compounds according to their elemental compositions such as the *Merck Index* and the chemical index within the *Chemical Abstracts*. The *NIST05 Mass Spectral Database* has proven to be very effective for the purpose of identifying candidate compounds in an unknown search.

The *Formula Search* feature of *MS Search* generates a list of compounds matching the entered formula. Compounds in the list that have no synonyms and/or are not included in other databases (other Database field in the records of the search results) are eliminated from consideration. The remaining compounds can be considered on a tentative basis.

The structures of the compounds in the tentative list of analytes are examined to determine the number of active hydrogens that are present on the compound. If active hydrogens are present on the analyte, the hydrogen/deuterium exchange analysis is performed. In this case, the tentative analytes were found to have active

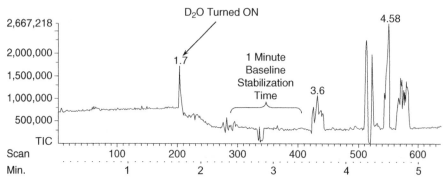

Figure 13.3. TIC of a data file showing the fluctuations that occur in the baseline as a result of the D_2O saturation of the sample gap.

hydrogens ($-OH$, $-NH_2$, or $-NRH$ groups). This procedure involving the introduction of D_2O in the sample gap in order to instigate H/D exchange is described elsewhere [7]. When the nebulizing gas is turned on and D_2O begins to fill the sample gap, a change in the baseline occurs. In Figure 13.3, notice that when the D_2O is turned on, the baseline drops. Once the D_2O nebulizer is turned on, the baseline should be allowed to stabilize for approximately one minute before any sample is introduced into the sample gap. In the presence of D_2O, the background peaks represent ions that contain active hydrogen atoms that undergo H/D exchange causing the appearance of the real-time mass spectrum to change further indicating the D_2O saturation of the sample gap. In order to properly calibrate the acquired data file, PEG must be analyzed at 200 °C in the absence of D_2O; this can be done after all samples have been analyzed. The H/D exchange is only valid for samples analyzed using the positive-ion detection mode.

The number of H/D exchanges that are observed with the analyte is determined. The numbers of H/D exchanges can aid in narrowing the list of tentative compounds; the number of active hydrogen atoms, n, as well as the proton associated with the protonated molecule is predicted to exchange with deuterium to form $n + 1$ ions.

Further confirmation of the unknown's identity can be obtained by the use of in-source collisionally activated dissociation. Once tentative compounds are selected, an authentic sample of each compound is obtained. These samples and the unknowns are analyzed with Orifice-1 of the *AccuTOF* mass spectrometer operated at increasingly higher voltages (40 V, 50 V, and 60 V). This can be done using a switching function method or individual data acquisitions using different Orifice-1 voltages. In order to properly calibrate the acquired data file, PEG must be analyzed at 200 °C. Because Orifice-1 is operated at higher voltages, which affects the PEG polymers, the calibration may not be as accurate.

Using the mass spectrometer's data analysis software, a comparison is made of the mass spectra acquired from the samples analyzed with Orifice-1 operated at 20 V and at higher Orifice-1 voltages. "New peaks" in the spectra obtained using higher voltages are assigned and are assumed to represent in-source collisionally activated dissociation (CAD) fragment ions. Elemental Compositions software is used to determine the elemental compositions of the ions represented by these "new mass spectral peaks." The elemental compositions of these ions are compared with elemental compositions of the suspected analyte to see if logical losses from the protonated molecule of the tentative compounds can be confirmed. Further confirmation is made by the analysis of the authentic samples with the presence of D_2O in order to confirm identical H/D exchanges.

13.4.2   Results and Discussion

In earlier studies, the DART heater was operated at 200 °C. When this temperature was used with the filter paper, degradation of the filter paper occurred as indicated by a slight brown mark apparent only after introduction into the heated DART helium stream. The mass spectrum of the singed portion of the filter paper exhibited new mass spectral peaks not observed at lower temperatures; the presence of new mass spectral peaks may indicate that the filter paper was burning at 200 °C. The DART heater temperature was lowered to 100 °C for all subsequent filter paper analyses in order to reduce the degradation of the matrix and the unknown.

In order to identify the unknown captured on the filter paper, an analysis of both an unused blank filter and a used filter was performed. The acquired mass spectrum of each is shown in Figure 13.4 presented in the centroid mode. No mass spectral peaks above 5% relative intensity of the base peak were detected beyond m/z 300 in either spectrum; therefore, each spectrum is displayed with the spectral range of m/z 50 to 300.

First inspection of the mass spectrum of the unused blank filter (Figure 13.4A) indicates that the mass spectral peaks at nominal m/z 114 and 229 are both base peaks. However, inspection of the tabular data reveals that the peak at m/z 114 is actually the base peak. It should be noted that the intensity of this base peak is ~50 K counts, which is a factor of five less than the intensity of the base peak in the spectrum obtained from the used filter (Figure 13.4B). The other prominent mass spectral peaks in the spectrum of the blank filter are at nominal m/z 246, 217, 202, 100, 74, and 57.

The mass spectrum of the used filter (Figure 13.4B) is surprisingly clean, exhibiting only four mass spectral peaks. Each of the four peaks has significantly different masses as compared to the mass spectral peaks observed in the mass spectrum of the unused blank filter (the blank); therefore, these four peaks are associated with substances captured on the filter paper. The base peak of the mass spectrum of the used filter (Figure 13.4B) is at m/z 135.0928, which could represent a protonated molecule, $[M + H]^+$; the mass spectral peak at m/z 269.1810 with a 65% relative intensity has the correct isotope intensities to represent a protonated dimer, $[2M + H]^+$, of this compound. The high intensity of the mass spectral peak at m/z 269.1810 suggests that the compound represented by this ion and the ion with m/z 135.0928 is at high concentrations in the interface. The high concentration of the ion with nominal m/z 135 in the sample interface may suggest that there may be more than one pathway by which this ion forms—a protonated molecule and by the loss of a

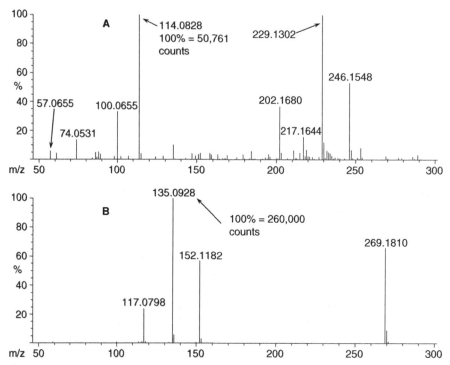

Figure 13.4. Mass spectra of filter paper presented in the centroid mode. A) Spectrum of unused blank filter paper with prominent mass spectral peaks labeled with m/z values. The base peak is at nominal m/z 114 at 50,761 counts. B) Spectrum of used filter paper with mass spectral peaks associated with the unknown labeled with m/z values. The base peak is at nominal m/z 135 at 260,000 counts.

molecule of ammonia, $[M + H - NH_3]^+$, from the ion represented by the mass spectral peak at m/z 152.1182 with a 55% relative intensity. The mass spectral peak at m/z 117.0798 may represent an ion that results from a loss of a molecule of water, $[M + H - H_2O]^+$, from the ion with m/z 135.0928; the presence of this mass spectral peak indicates that the unknown may contain at least one hydroxyl group.

An in-source CAD analysis was performed with Orifice-1 operated at 40 V on the used filter to confirm the assumption that all four mass spectral peaks are associated with one substance captured on the filter paper. The mass spectrum (presented in the centroid mode) of the unknown captured on filter paper produced by this fragmentation method is shown in Figure 13.5. The protonated molecule is represented by the mass spectral peak at m/z 135.0967 with an 80% relative intensity; the high intensity of this peak remaining at a higher Orifice-1 voltage supports the proposition that the mass spectral peak at m/z 269.1911 represents a protonated dimer of this compound. The presence of a mass spectral peak at m/z 117.0853 with a 45% relative intensity indicates that at least one hydroxyl group is present; this peak represents an ion that is formed when the ion with m/z 135.0967 loses a water molecule. The base peak of the mass spectrum is at m/z 59.0471.

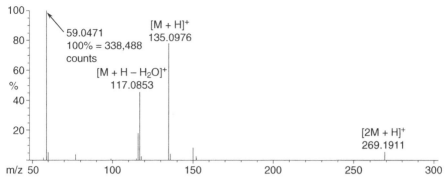

Figure 13.5.   In-source CAD fragmentation mass spectrum of unknown on used filter paper. Spectrum is presented in centroid mode. The base peak of the spectrum represents a fragment ion of the protonated molecule.

The results of the in-source CAD analysis indicates that one substance, the unknown, is captured on the filter paper and the unknown's protonated molecule is represented by the mass spectral peak detected at m/z 135.0967. Using accurate mass measurements, the elemental composition was determined to be $C_6H_{15}O_3$ utilizing the *Elemental Compositions* software Version 3.1 (available in *MS Tools* software). The determined elemental composition's exact mass was within ±3 millimass units (mmu) of the measured mass, a difference of only −0.00274 u, which is adequate for an unambiguous elemental composition at this mass. This elemental composition was confirmed through the use of isotope peak-intensity ratios. A *Formula Search* of $C_6H_{14}O_3$ using the *NIST05 Mass Spectral Database* produced 13 candidate compounds. Compounds that had no synonyms or commercial names and were not common to multiple databases were eliminated from consideration. Because the in-source CAD fragmentation spectrum of the unknown confirmed the presence of at least one hydroxyl group contained in the unknown, any compound without a hydroxyl group was also eliminated from consideration. Two constitutional isomers that are commercially common and logically related to the possible origin of the unknown compound were good candidates for the unknown's identity; one contained a single hydroxyl group (diethylene glycol monoethyl ether [DEGE], a common solvent and humectant) and the other contained two hydroxyl groups (dipropylene glycol [DPG], a common component of plasticizers, flavors, and fragrances).

Unfortunately, the in-source CAD spectrum, neither through the nominal nor the accurate mass of the fragment ions, allowed for differentiation between the two candidate compounds because of their elemental and structural similarities. At the time of the initial analyses, authentic samples of the two analytes were not available to determine if the spectrum produced by in-source CAD was the same for both compounds. The dilemma was resolved using in-source H/D exchange on the DART. Figure 13.6 shows the mass spectrum of the unknown on the used filter analyzed in the presence of D_2O to facilitate H/D exchange of any active hydrogen atoms contained in the unknown. The mass spectrum is expanded in order to focus on the number of H/D exchanges that occurred on the protonated molecule. The unknown

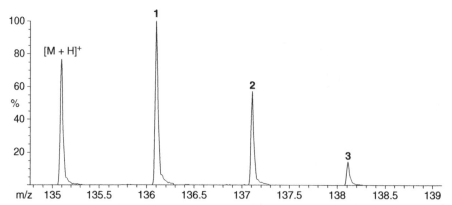

Figure 13.6.   Mass spectrum of the unknown on used filter paper in the presence of D_2O. Spectrum is presented in the profile mode with the protonated molecule region of the spectrum expanded. The protonated molecule is labeled with $[M + H]^+$. The mass spectral peaks due to H/D exchange are labeled with the number of hydrogen atoms exchanged with deuterium.

analyte's mass spectrum obtained through H/D exchange exhibited a cluster of peaks representing the protonated molecule and three other peaks representing an $n + 1$ exchange where n was 2. This meant that the analyte had to have two active hydrogen atoms; i.e., the analyte had to be DPG.

Authentic samples of both DPG and DEGE were obtained. The in-source CAD fragmentation spectra of these authentic samples produced a mass spectral peak representing the protonated molecule and the loss of water from the protonated molecule, as expected (Figures 13.7A and 13.7B, respectively). The spectra also showed unique fragment ions for DPG (nominal m/z 59) and DEGE (nominal m/z 73 and 89). The in-source CAD fragmentation spectra of the two authentic compounds were different enough to differentiate between the constitutional isomers. The in-source CAD mass spectrum of the unknown matched the in-source CAD mass spectrum of DPG obtained at the same Orifice-1 voltage.

Any ambiguity that may have resulted from the possibility of other analytes being present was eliminated through the use of H/D exchange. The mass spectrum obtained through H/D exchange analysis for both compounds is shown in Figure 13.8. The DEGE spectrum obtained with H/D exchange shows a peak representing the protonated molecule at m/z 135.1024 as well as two mass spectral peaks representing two consecutive H/D exchanges. The DPG mass spectrum obtained using H/D exchange exhibits the protonated molecule peak at m/z 135.0991 as well as three mass spectral peaks representing three consecutive H/D exchanges. Each compound's spectrum obtained with H/D exchange was compared to the H/D exchange spectrum of the unknown compound captured on filter paper. Both the unknown and the DPG spectra exhibit a cluster of peaks representing the protonated molecule and three consecutive H/D exchanges. Due to the similarity of peak clusters between the spectra obtained using the H/D exchange for the DPG and the unknown associated with the filters, the identity of the unknown was confirmed (by

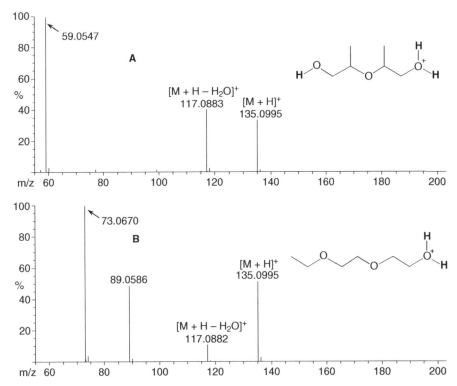

Figure 13.7.   Fragmentation mass spectra of DPG (A) and DEGE (B). Mass spectra acquired with Orifice-1 operated at 40 V presented in the centroid mode. Each peak representing the protonated molecule is labeled with $[M + H]^+$.

cross-correlation of results from accurate mass measurement, accurate isotope peak intensities, in-source CAD, and H/D exchange) to be DPG, dipropylene glycol.

Once the unknown was identified as DPG, further analyses were performed in order to understand the presence of the mass spectral peak at nominal m/z 152. Based on measured accurate mass and accurate isotope peak intensities, the elemental composition was consistent with an ammonium adduct on DPG, $[M + NH_4]^+$. The origin of the ammonia may be due to the use of the reagent in the laboratory. When DPG was analyzed with ammonia saturated in the DART sample gap in order to force the production of $[M + NH_4]^+$, the peak at nominal m/z 152 became the base peak confirming that the mass spectral peak at nominal m/z 152 is related to DPG. Also, assuming the peak at nominal m/z 152 represents an ammonium adduct, $[M + NH_4]^+$, all of the hydrogen atoms associated with the ammonium ion are considered active and will therefore exchange with deuterium in the presence of D_2O; therefore, this ion will undergo six H/D exchanges. Upon inspecting the mass spectrum acquired when the used filter was analyzed in the presence of D_2O, the cluster of peaks beginning with the peak at nominal m/z 152 shows six peaks consistent with six H/D exchanges (Figure 13.9). The results of these two experiments further demonstrate that only one substance was captured on the used filter paper, DPG.

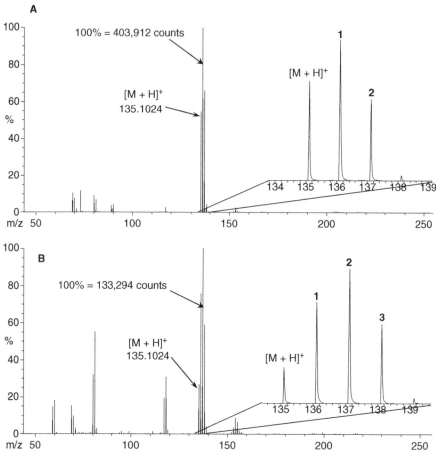

Figure 13.8. Mass spectra of candidate compounds DEGE and DPG in the presence of D$_2$O. A) Spectrum of DEGE with the peak cluster beginning at nominal *m/z* 135 enlarged in order to see the presence of a peak representing [M + H]$^+$ and two consecutive H/D exchanges. B) Spectrum of DPG with the peak cluster beginning at nominal *m/z* 135 enlarged in order to see the presence of a peak representing [M + H]$^+$ and three consecutive H/D exchanges. Each mass spectrum is presented in the profile mode.

13.5 CONCLUSIONS

Because of the ability to provide unambiguous elemental compositions from accurate mass measurements, the time-of-flight mass spectrometer is ideally suited for use with devices such as the DART. Using the TOF's accurate mass measurement, in-source CAD, and in situ H/D exchange, these instruments can be used to identify many unknowns. The method described here is dependent on an electron ionization mass spectrum of the analyte being in the *NIST05 Mass Spectral Database*; however, with the spectra of 163,198 compounds in the current version of the Database

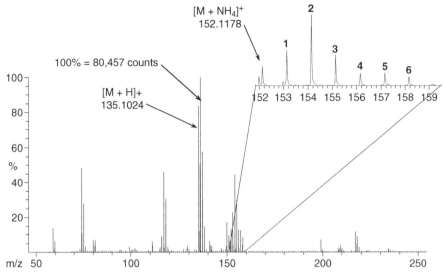

Figure 13.9.   Mass spectrum of used filter analyzed in the presence of D₂O. The peak cluster beginning at nominal *m/z* 152 is enlarged in order to see the presence of a peak representing [M + NH₄]⁺ and six consecutive H/D exchanges.

(NIST05), and 28,297 of these listed in other databases, the likelihood of an identification is good. As was stated above, there are other sources of listings of compounds according to their elemental compositions such as the *Merck Index* and the chemical index within the *Chemical Abstracts* that could be used for the same purposes.

REFERENCES

1. Sparkman, O.D., MassSpectrometry PittCon® 97, *J. Am. Soc. Mass Spectrom.*, **1997**, 8, 573–579.
2. Gross, M.L., Accurate masses for structure confirmation, editorial, *J. Am. Soc. Mass Spectrom.*, **1994**, 5, 57.
3. Cody, R.R.; Laramée, J.A.; Durst, H.D., Versatile new ion source for the analysis of materials in open air under ambient conditions, *Anal. Chem.*, **2005**, 77, 2297–2302.
4. Sparkman, O.D., Additional real time structural information from DART mass spectrometry, Presented at PittCon 2007 (March), Chicago, IL.
5. Vanderford, B.J.; Pearson, R.A.; Cody, R.B.; Rexing, D.J.; Snyder, S.A., Determination of an unknown contaminant using LC/MS/MS, *in Liquid Chromatography/Mass Spectrometry, MS/MS and Time-of-Flight MS: Analysis of Emerging Contaminants* Ferrer I.; Thurman E.M., Eds. ACS Symposium Series 850. American Chemical Society, Washington, DC, **2003**.
6. Greiner, W., *Quantum Mechanics: An Introduction*, Second corrected edition; Springer: Berlin, **1994**.
7. Vail, T.; Jones, P.R.; Sparkman, O.D., Rapid and unambiguous identification of melamine in contaminated pet food based on mass spectrometry with four degrees of confirmation. *J. Anal. Toxicol.*, **2007**, 31, 304–312.

REFERENCES

CHAPTER *14*

COMPARISON OF LC/TOF-MS AND LC/MS-MS FOR THE ANALYSIS OF 100 PESTICIDES IN FOOD: FINDING THE "CROSSOVER POINT"

E. Michael Thurman and Imma Ferrer

Center for Environmental Mass Spectrometry, Department of Civil, Environmental and Architectural Engineering, University of Colorado, Boulder, Colorado

Jerry A. Zweigenbaum and Paul A. Zavitsanos

Agilent Technologies, Inc., Wilmington, Delaware

ANALYTICAL METHODS for determining the presence of a group of 100 pesticides in vegetable and fruit samples were developed using a time-of-flight mass spectrometer (LC/TOF-MS) and a triple quadrupole mass spectrometer (LC/MS/MS) and the two methods were compared. The sensitivity of the LC/TOF-MS was determined in food matrices of orange, tomato, and green pepper using accurate mass in full scan. For the LC/MS/MS, two transitions per parent compound were monitored in a single chromatographic run containing two time segments. The sensitivity obtained meets the maximum residue levels (MRLs) established by the European Union regulation for food monitoring programs. Likewise, this study is a valuable indicator of the potential of both the LC/TOF-MS and the LC/MS/MS for routine quantitative multi-residue analysis of pesticides in vegetables and fruits and discusses the limit-of-detection or "crossover point," which is the point where the sensitivity of the LC/TOF-MS and the LC/MS/MS are approximately equal for 50% of the compounds. Beyond this point the LC/TOF-MS becomes more practical for pesticide screening as the sensitivity is greater for the majority of compounds.

Liquid Chromatography Time-of-Flight Mass Spectrometry: Principles, Tools, and Applications for Accurate Mass Analysis, Edited by Imma Ferrer and E. Michael Thurman
Copyright © 2009 John Wiley & Sons, Inc.

14.1 INTRODUCTION

The European Union (EU) has set directives for pesticides at low levels in fruits and vegetables intended for baby food with an MRL of 0.01 mg/kg, which is also applicable for all pesticides and compounds without a stated regulation. The low MRLs have encouraged the development of more sensitive analytical methods to meet the requirements in complex samples as well as a long list of potential pesticide candidates that spurs continued methods development for ultra sensitive methods. In this sense, liquid-chromatography tandem–mass spectrometry (LC/MS/MS) with triple quadrupole in multiple reaction monitoring (MRM) mode has become, so far, the most widely used technique for the monitoring and quantitation of pesticides in food, as reported extensively in the literature [1–3]. On the other hand, high-resolving power mass spectrometric techniques, such as time-of-flight mass spectrometry (LC/TOF-MS), have been applied recently for screening purposes, especially with the goal of screening hundreds of pesticides in food samples using database approaches [4–6]. Thus, an important question arises about which mass spectrometric instruments one should choose based on the number of compounds being analyzed and the detection limit sought.

This chapter is one of the first of its kind to examine and compare a LC/TOF-MS and a LC/MS/MS from the same manufacturer for screening of pesticides in food; thus, the comparison is considered valid. The advantage of the same manufacturer means that the electrospray source is equal in both cases, the same HPLC column is used, as well as much of the basic electronics in up-front lenses and computing power. The value of this comparison is that one can determine the "crossover point," or the number of compounds where the sensitivity of LC/TOF-MS and LC/MS/MS would be equal for 50% of the compounds under study. Because of the need to screen hundreds of pesticides in food, this comparison is a valid and important question to answer as those in the food monitoring business try to extend the number of pesticides to hundreds of compounds or more [1, 2, 4–6].

14.2 EXPERIMENTAL

14.2.1 Sample Preparation

Pesticide analytical standards were purchased from ChemService (West Chester, PA) and from Sigma Aldrich (St. Louis, MO). Individual pesticide stock solutions (1000 μg/mL) were prepared in pure acetonitrile or methanol depending on the solubility of each individual compound, and stored at −18 °C. From these solutions, working standard solutions were prepared by dilution with acetonitrile and water.

Vegetable samples were obtained from the local markets. "Blank" vegetable and fruit extracts were used to prepare the matrix-matched standards for validation purposes. In this way, two types of vegetables and one fruit (green peppers, tomatoes, and oranges) were extracted using the ethyl acetate method [7]. The vegetable extracts were spiked with the mix of standards at different concentrations (ranging from 0.1 to 100 µg/kg) and subsequently analyzed by LC/TOF-MS and by LC/MS/MS (QqQ).

14.2.2 LC/TOF-MS Conditions

LC Pumps were Agilent 1100 binary pumps with an injection volume of 50 µL with standard Agilent 1100 automated liquid sampler. The column was a ZORBAX Eclipse® XDB 4.6 × 150 mm C-8, 5-micron packing. The mobile phase was A = acetonitrile and B = 0.1% formic acid in water. The gradient began with 5 minutes isocratic at 10% A followed by a linear gradient to 100% A in 25 minutes at a flow rate of 0.6 mL/min. The mass spectrometer consisted of a model LC/MSD TOF with electrospray source operated in positive ESI with a capillary voltage of 4000 V, nebulizer pressure of 40 psig, drying gas 9 L/min, and gas temperature of 300 °C. The fragmentor was set at 190 V, skimmer 60 V, Oct DC1 37.5 V, and OCT RF V at 250 V. The reference masses were m/z 121.0509 and 922.0098, resolution: 9500 ± 500 @ m/z 922.0098, mass range (m/z) 50–1000; reference A sprayer 2 is operated at a constant flow rate during the run.

14.2.3 LC/MS-MS Conditions

The liquid chromatographic conditions were identical for the triple quadrupole analysis, except a 10 µL sample was injected onto an Agilent SB-C-18, 4.6 mm × 150 mm, 1.8 µm column and an Agilent Model 1200 LC was used. See conditions above. The mass spectrometer was the Agilent (Santa Clara, CA) model 6410 triple quadrupole, with the electrospray positive source operated with a nebulizer pressure of 40 psig, drying gas of 9 L/min, capillary voltage of 4000 V, drying gas temperature of 350 °C, fragmentor voltage of 70 V, and collision energies from 5 to 25 volts depending on the compound and its transitions. The dwell time was 15 milliseconds for each compound, which includes the dwell time of 10 ms and the 5 ms pause time between transitions [1].

14.3 RESULTS AND DISCUSSION

14.3.1 Limits of Detection: Triple Quadrupole

A study of the optimal MRM transitions for 100 pesticides was carried out by injecting groups of analytes (around 10 analytes in one chromatographic run) at a concentration level of 10 µg/mL. Various fragmentation voltages and collision energies were applied to the compounds under study in order to obtain the limits-of-detection reported in this study and published previously by Ferrer et al. [1]. The optimum energies were those that gave the best sensitivity for the main product ion and left about 10% of parent compound in the spectra, and they were selected as the optimum collision energies [1] for the data in Table 14.1. Two transitions were found to develop a comprehensive analysis method for these pesticides using QqQ. The

TABLE 14.1 Comparison of LODs for QQQ and LC/TOF-MS for 100 Pesticides Ranked by Alphabetical Order with Concentrations in μg/kg Analyzed in Food Samples [1, 8]. The QQQ Used Two Windows of 50 Transitions (MRMs) Each

Compound Name	Ret. Time (min)	Protonated Molecule [M + H]⁺	Product Ion	QqQ LOD's (μg/kg)	LC/TOF-MS LOD's (μg)/kg
Acetamiprid	12.2	233	126	0.3	3
Acetochlor	23.1	270	224	0.8	0.6
Alachlor	23.1	270	238	0.8	3
Aldicarb	14.3	116	89	2	5
Aldicarb sulfone	7.9	223	76	5	3
Aldicarb sulfoxide	6.1	207	89	2	4
Atrazine	17.5	216	174	0.4	0.5
Azoxystrobin	21.3	404	372	0.3	0.1
Benalaxyl	24.4	326	148	0.5	0.04
Bendiocarb	16.5	224	109	1.0	5.0
Bensulfuron methyl	19.0	411	149	0.4	1.0
Bromoxynil	17.9	278	199	40	20
Bromoconazole	21.5	376	159	1.0	0.3
Buprofezin	26.6	306	201	0.7	0.6
Butylate	27.7	218	57	5.0	3.0
Carbaryl	17.4	202	145	10	3.0
Carbendazim	7.1	192	160	0.5	0.8
Carbetamide	13.9	237	118	0.5	1.0
Carbofuran	16.6	222	165	0.9	4.0
Chlorfenvinphos	23.7	359	155	2.0	0.2
Chlorotoluron	16.8	213	72	0.3	1.0
Chlorpyrifos methyl	25.9	322	125	10	30
Cyanazine	15.3	241	214	2.0	2.0
Cyproconazole	20.3	292	70	0.5	1.0
Cyromazine	3.4	167	85	10	9.0
Deethylatrazine	11.2	188	146	1.0	2.0
Deethylterbuthytazine	15.4	202	146	0.8	1.5
Deisopropylatrazine	8.7	174	96	4.0	2.0
Diazinon	25.3	305	169	0.3	0.05
Dichlorvos	15.4	221	109	5.0	0.5
Difeconazole	24.7	406	251	0.3	0.5
Difenoxuron	18	287	72	0.6	0.4
Diflubenzuron	22.3	311	158	6.0	12
Dimethenamide	21.2	276	244	0.4	1.0
Dimethoate	11.8	230	199	0.7	1.5
Dimethomorph	19.2	388	301	0.6	4.0
Diuron	17.1	233	72	0.8	0.6
Ethiofencarb	17.9	226	107	0.7	4.0
Fenamiphos	20.8	304	217	0.6	0.1
Fenuron	11.2	165	72	1.5	10
Flufenacet	23	364	152	0.5	3.0
Flufenoxuron	27.6	489	158	5.0	6.0

TABLE 14.1. *(Continued)*

Compound Name	Ret. Time (min)	Protonated Molecule [M + H]⁺	Product Ion	QqQ LOD's (μg/kg)	LC/TOF-MS LOD's (μg)/kg
Fluometuron	17.9	233	72	1.0	5.0
Fluroxpyr	14.9	255	209	10	20.0
Hexaflumuron	25.1	461	158	7.0	8.0
Hydroxyatrazine	8.1	198	156	4.0	0.4
Imazalil	18.5	297	159	10	0.3
Imazapyr	9.2	262	217	0.7	5.0
Imazaquin	15.4	312	199	0.6	0.7
Imidacloprid	11.4	256	175	4.0	2.0
Ioxynil	19.6	372	118	20	15
Iprodione	22.6	330	245	12	4.0
Irgarol 1051	19.2	254	198	0.8	0.1
Irgarol metabolite	13.6	214	158	1.2	0.5
Isofenphos	26.4	346	217	1.0	2.0
Isoproturon	17.7	207	72	1.3	0.7
Lenacil	15.5	235	153	8.0	9.0
Linuron	20.7	249	160	1.0	2.0
Lufenuron	26.8	511	158	3.0	9.0
Malathion	22.7	331	99	0.8	1.5
Mebendazole	14.8	296	264	0.6	0.8
Metalaxyl	17.7	280	192	1.0	0.2
Metamitron	10.6	203	175	0.9	3.0
Methidathion	20.8	303	85	0.7	15
Methiocarb	20.4	226	121	0.8	0.7
Methiocarb sulfone	13.2	258	122	30	9.0
Methomyl	8.6	163	88	0.8	2.0
Metolachlor	23.2	285	252	0.4	0.8
Metolcarb	15.3	166	109	2.0	12
Metribuzin	15.9	215	187	1.0	0.6
Molinate	22.2	188	126	2.0	1.5
Monuron	14.9	199	72	1.5	0.7
Nicosulfuron	13.7	411	182	0.8	0.8
Nitenpyram	11	271	225	0.7	0.2
Oxadixyl	14.9	279	219	5.0	14
Parathion ethyl	24.6	292	236	5.0	17
Pendimethalin	28.5	282	212	4.0	11
Phosmet	21.2	318	160	6.0	0.9
Prochloraz	23.2	376	308	5.0	0.7
Profenofos	26.6	373	303	5.0	1.0
Promecarb	20.9	208	109	0.7	3.0
Prometon	14	226	142	2.0	1.0
Prometryn	18.3	242	158	0.9	0.3
Propachlor	19.1	212	170	1.0	0.5
Propanil	19.8	218	127	0.8	0.7

(Continued)

TABLE 14.1. *(Continued)*

Compound Name	Ret. Time (min)	Protonated Molecule [M + H]⁺	Product Ion	QqQ LOD's (µg/kg)	LC/TOF-MS LOD's (µg)/kg
Propiconazole	23.7	342	159	0.7	0.3
Prosulfocarb	27.1	252	91	0.6	2.0
Simazine	14.9	202	132	0.7	0.4
Spiromesifen	30.1	371	273	7.0	120
Sulfosulfuron	18.4	471	211	0.8	2.0
Teflubenzuron	25.6	381	158	9.0	35
Terbuthylazine	20.5	230	174	0.3	0.4
Terbutryn	18.6	242	186	1.0	0.3
Thiabendazole	7.8	202	175	6.0	5.0
Thiacloprid	14.0	253	126	2.0	1.5
Thiocyclam	6.3	182	137	50	5.0
Triazophos	22.9	314	162	0.6	1.0
Triclocarban	25.2	315	162	2.0	12
Trifloxystrobin	26.1	409	186	0.4	2.0
Triflumizole	24.9	346	278	3.0	3.0

stronger MRM transition was used for quantitation and the second MRM transition was used for confirmation [1].

One of the most sensitive pesticides was diazinon at 3 pg on column with a detection limit in food of 0.3 µg/kg. The median LOD was 10 pg on column or a LOD of 3 µg/kg. The least sensitive compounds had a LOD of 30–50 picograms and included compounds such as imazalil, which were quite sensitive by LC/TOF-MS. This apparent contradiction that some compounds with the same source (i.e., Agilent electrospray positive source) could be more sensitive by LC/TOF-MS by as much as 5 times was because of the requirement of a second transition for the QqQ analysis. This second transition could sometimes be weak and insensitive, while the first transition was strong. Thus, the second confirmation transition was responsible for raising the LOD, as in the case for imazalil, to a value that was considerably greater than the LC/TOF-MS. This was an unexpected result but not necessarily unusual. Finally, the number of transitions that were required to analyze 100 pesticides was thus a total of 200 transitions and this value was used for the comparison with the LC/TOF-MS.

14.3.2   Limits of Detection: LC/TOF-MS

Likewise, the LC/TOF-MS was optimized for fragmentor voltage and the same chromatographic conditions were used as the QqQ so that results were comparable. Results for the LOD and the protonated molecule for the LC/TOF-MS are shown in Table 14.1 as previously published by Ferrer and Thurman [8]. The LODs varied

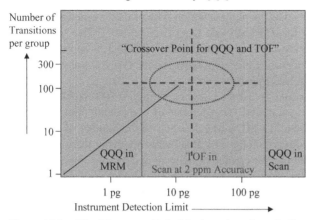

Figure 14.1.   The "Crossover Point" is the point where QqQ and LC/TOF-MS have similar limits of detection for pesticide screening for 50% of the compounds.

from 1 picogram for diazinon to as much as 300 picograms for spiromesifen, an insensitive compound. The median LOD was 20 picograms on column. The LOD was determined so that not only did the protonated molecule need to give the correct accurate mass with a 5-ppm mass accuracy but also the A+1 isotope of the protonated molecule. The 5-ppm mass accuracy was used because this is the value commonly accepted for the minimum mass accuracy for the determination of an unknown in analytical analysis [9]. The compounds that were considerably more sensitive by LC/TOF-MS than by LC/QqQ are benalaxyl, chlorfenvinphos, diazinon, dichlorvos, fenamiphos, hydroxyatrazine, imazalil, irgarol 1051, metalaxyl, phosmet, prochloraz, profenofos, and thiocyclam. The reason is as explained above for imazalil, that by requiring the second confirmation transition it is possible to raise the LOD by a substantial amount for the triple quadrupole.

Thus, in general the LC/QqQ is approximately 5 to 50 times more sensitive than the LC/TOF-MS when both instruments are optimized for best performance and analysis of pesticides. The majority of the compounds (90%) were more sensitive by triple quadrupole MS (QqQ), which varied from 2 to 500 times more sensitive than LC/TOF-MS. The median sensitivity difference (50% of the compounds) was 5 times greater by QqQ than by LC/TOF-MS. The LC/TOF-MS was more sensitive on 10% of the compounds (Table 14.1). This preliminary comparison of LODs suggest that if the number of MRM transitions of the QqQ is increased from 50 to 100 transitions per window, which forces the dwell time of the QqQ to decrease to its limit of 10 milliseconds, one would reach equal LODs for the two instruments (Figure 14.1 and Table 14.1). But this comparison is based on the optimum dwell time for the QqQ. What happens if this optimum dwell time is shortened and the number of transitions is increased? Is there a point at which the two instruments share the same LOD and at what concentrations is this?

14.3.3 The Crossover Point

The "crossover point" is defined as the point at which 50% of the screened pesticides would have equal LODs by LC/TOF-MS and QqQ (Figure 14.1). The "crossover point," which is shown in Figure 14.1 and derived from Table 14.1, is approximately 20 pg for LC/TOF-MS, which corresponds to a QqQ segment of 200 transitions per window or 100 analytes with two transitions each.

Thus, 200 transitions would give a screening capability of 100 compounds per window, with 2 MRMs per compound. The number of compounds that could be screened by QqQ compared with LC/TOF-MS depends on the number of windows that can be inserted in a chromatographic run. Practically speaking there is a limit to the number of windows that may be created because of two reasons. First involves the possibilities of common product ions and something called "cross talk," which is when product ions are left in the collision cell and found in the next transition. The second reason is because of the overlap of retention times of the pesticides, most of which elute from the C-8 column between 17 and 30 minutes, with more than 50% of the compounds occurring in a short 5-minute window of 18–23 minutes, which is caused by the similar affinity of many of the pesticides for the C-8 stationary phase (Table 14.1) or for a C-18 column, for that matter.

There are chromatographic challenges that must be met as well as mass spectrometery challenges for screening hundreds of pesticides by QqQ. For all practical purposes, we set the cross over point for LC/QqQ and LC/TOF-MS at approximately two segments at the critical window time of 17–30 minutes with a total of 200 compounds with 400 total transitions for pesticides. Above this value is not only more practical but more useful to use LC/TOF-MS. Having said this, we have been able to reach a total of 300 pesticides in a single analysis using LC/QqQ [10] but for such large analysis sets it is more practical to use LC/TOF-MS or LC/Q/TOF-MS.

To prove this point, Figure 14.2 shows the analysis of 300 pesticides by LC/Q-TOF/MS using a similar C-8 chromatographic column. The chromatogram is quite complicated yet it is easily to do a scheduled MS/MS analysis of a pesticide, in this case, atrazine, and to generate not only the accurate mass of the precursor ion (in this case the protonated molecule) but also the accurate mass of all the fragment ions. Figure 14.3 shows the pathway that furthermore can be determined from such an accurate mass analysis. This example shows atrazine with 10 fragment ions. The accurate mass and combination of MS/MS capabilities allows definitive pathway analysis of fragment ions.

14.4 CONCLUSIONS

Our preliminary study of the "crossover point" between the QqQ and LC/TOF-MS is estimated at a LOD concentration of ~20 pg on column for the LC/TOF-MS, which is approximately 200 transitions per window by QqQ, or approximately 100 compounds per window. This result suggests that as many as 200–300 pesticides may

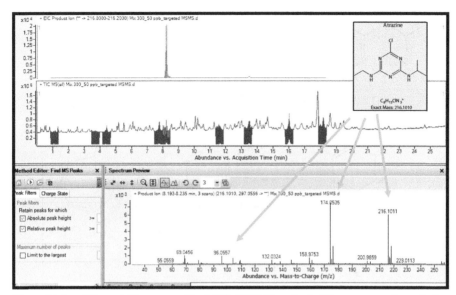

Figure 14.2.   300 pesticides targeted by LC/Q-TOFMS analysis. The spectrum shown is for atrazine, from a mixture of pesticides.

be screened by QqQ before reaching a method "crossover point." However, it must be pointed out that because a majority of pesticides elute in a narrow window of time (of medium hydrophobicity on C-8) it may be practically very difficult to break the chromatographic run into multiple MRM transition windows. Because of this practical fact, it appears that LC/TOF-MS becomes a more viable pesticide screening-method than QqQ somewhere in the ~200–250 compound region. Furthermore, there is literally no limit to the number of compounds that may be screened by LC/TOF-MS because of its continuous full scan capability with accurate mass at 2-ppm, high resolving power (>10,000), and a median LOD of approximately 20 pg on column for more than 50% of the pesticides studied. For both a screening and confirmation tool in a single instrument for large numbers of pesticides (>200–250 compounds), the use of LC/Q-TOFMS is shown. Here one can obtain not only accurate mass of the protonated molecule but also of the major fragment ions of the pesticide.

Thus, the triple quadrupole (QqQ) and the LC/TOF-MS are considered compatible instruments rather than competitive, with the QqQ more sensitive for pesticides when the number of pesticides is kept well below 200 compounds/method. LC/TOF-MS is best for screening of large numbers of pesticides, greater than 200–250 or more. Both instruments are capable of good linearity over three orders of magnitude, as we have discussed in previous papers [1, 5]. So both instruments are suitable for quantitative work on pesticides in vegetable samples.

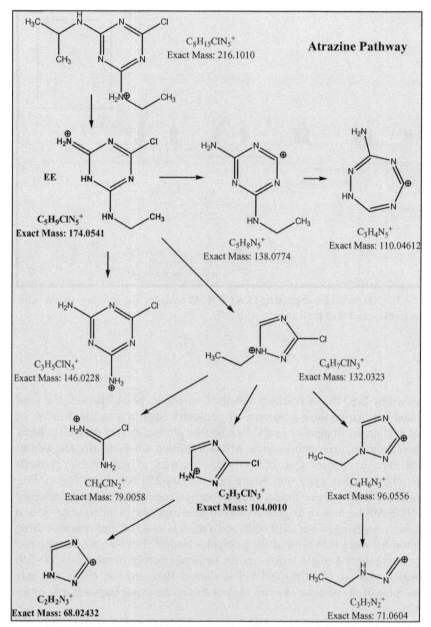

Figure 14.3. Pathway of degradation for atrazine using LC/Q-TOFMS analysis from 300 pesticide mixture shown in Figure 14.2.

REFERENCES

1. Ferrer, I.; Thurman, E.M.; Zweigenbaum, J.A., Screening and confirmation of a 100 pesticides in food samples by liquid chromatography/tandem mass spectrometry, *Rapid Commun. Mass Spectrom.*, **2007**, 21, 3869–3882.
2. Ferrer, I.; Thurman, E.M.; Fang, Y.; Zavitsanos, P.; Zweigenbaum, J.A., Multiresidue analysis of 100 pesticides in food samples by LC/triple quadrupole mass spectrometry, 2006, Agilent Technologies, Publication 5989-2209EN www.agilent.com/chem.
3. Hernandez, F.; Pozo, O.J.; Sancho, J.V.; Bijlsma, L.; Barreda, A.; Pitarch, E., Multiresidue liquid chromatography tandem mass spectrometry determination of 52 non gas chromatography amenable pesticides and metabolites in different food matrices, *J. Chromatogr. A*, **2007**, 1109, 242–252.
4. Ferrer, I.; Fernandez-Alba, A.; Zweigenbaum, J.A.; Thurman, E.M., Exact mass library for pesticides using a molecular feature database, *Rapid Commun. Mass Spectrom*, **2006**, 20, 3659–3668.
5. Thurman, E.M.; Ferrer, I.; Malato, O.; Fernandez-Alba, A., Feasibility of LC/TOFMS and elemental database searching as a spectral library for pesticides in food, *Food Additives and Contaminants*, **2006**, 23, 1169–1178.
6. Thurman, E.M.; Ferrer, I.; Zweigenbaum, J.A., Automated screening of 600 pesticides in food by LC/TOF-MS using a molecular feature database search, 2006, Agilent Technologies, Publication 5989-5496EN www.agilent.com/chem.
7. Pihlstrom, T.; Blomkvist, G.; Friman, P.; Pagard, U.; Osterdahl, B.G., Analysis of pesticide residues in fruit and vegetables with ethyl acetate extraction and using gas and liquid chromatography with tandem mass spectrometric detection, *Anal. Bioanal. Chem.*, **2007**, 389, 1773–1789.
8. Ferrer, I.; Thurman, E.M., Multi-residue method for the analysis of 101 pesticides and their degradates in food and water samples by liquid chromatography/time-of-flight mass spectrometry, *J. Chromatography A*, **2007**, 1175, 24–37.
9. Thurman, E.M.; Ferrer, I.; Zweigenbaum, J.A., High resolution and accurate mass analysis of xenobiotics in food, *Anal. Chem.*, **2006**, 78, 6703–6708.
10. Thurman, E.M.; Ferrer, I.; Zweigenbaum, J.A., Multiresidue analysis of 301 pesticides in food samples by LC/triple quadrupole mass spectrometry, **2008**, Agilent Technologies, Publication 5989-8614EN www.agilent.com/chem.

INDEX

Liquid Chromatography Time-of-Flight Mass Spectrometry: Principles, Tools, and Applications for Accurate Mass Analysis, Edited by Imma Ferrer and E. Michael Thurman
Copyright © 2009 John Wiley & Sons, Inc.

CHEMICAL ANALYSIS

A SERIES OF MONOGRAPHS ON ANALYTICAL CHEMISTRY
AND ITS APPLICATIONS

Series Editor
J. D. WINEFORDNER

Printed and bound by CPI Group (UK) Ltd, Croydon, CR0 4YY

16/04/2025

14658346-0005